Cybernetics and the
Management of Large Systems

Cybernetics and the Management of Large Systems

Proceedings of the
Second Annual Symposium
of the
American Society for Cybernetics

Edited by
Dr. Edmond M. Dewan
Data Sciences Laboratory
Air Force Cambridge Research Laboratories

SPARTAN BOOKS

NEW YORK • WASHINGTON

Library of Congress Catalog Card Number 71-83800
Standard Book Number 87671-123-9

Sole distributor in Great Britain, the British Commonwealth, and the
Continent of Europe:

MACMILLAN & CO. LTD.
Little Essex Street
London, W.C. 2

Printed in the United States of America.

Cybernetics and the
Management of Large Systems

Dedication

Dr. Nicholas E. Golovin (1912-1969), the author of "The Evaluative Function of Government," was in the most productive period of his career when he died unexpectedly on 27 April 1969.

Born in Odessa, Russia, he was educated in the United States, receiving AB and MA degrees from Columbia University in 1933 and 1936, and a PhD degree in physics from George Washington University in 1955.

An omnivorous reader, with a disciplined mind sharpened by experience, debate and creative thought, Golovin's holistic approach to social problems led him to write in a variety of fields, among them: research and development management, space science and technology, reliability engineering, creativeness in the sciences, problems of evaluation in education and the need for a fourth branch of government.

His wellsprings of energy seemed to be fed from an infinite reservoir. The heavier his burdens, the more buoyant his response. It was as though nature, in fashioning the man, forgot to incorporate any concept of weariness or limit to his capacity.

He chose to analyze nothing less than the universe of knowledge and experience, and has earned distinction as one of the fathers of social systems theory.

Generous in debate, possessing a keen intellect, serving his country with devotion and his friends with warm understanding, we dedicate this volume to

NICHOLAS E. GOLOVIN

HAROLD K. HUGHES
State University of New York
Potsdam, New York

Contents

III. DIRECT APPLICATION OF CYBERNETIC OR SYSTEM TECHNIQUES

The Participants

DR. HAROLD K. HUGHES
Vice President for Academic Affairs
State University of New York
Potsdam, New York

DR. DONALD O. WALTER
Brain Research Institute
University of California
Los Angeles, California

DR. STEPHEN L. SHERWOOD
Stanford University Medical School *and*
Veterans Administration Hospital
Palo Alto, California

DR. THOMAS H. NAYLOR
Professor of Economics
Duke University *and*
Carolina Population Center
Durham, North Carolina

DR. JOHN G. KEMENY
Director of Software Development *and*
Professor of Mathematics
Dartmouth College
Hanover, New Hampshire

MR. EARL C. JOSEPH
Chief Scientist
UNIVAC
St. Paul, Minnesota

MR. CHARLES J. PURCELL
Advanced Development Laboratories
Control Data Corporation
Minneapolis, Minnesota

DR. ARTHUR G. ANDERSON
Director of Research
International Business Machines Corporation
Yorktown Heights, New York

DR. MICHAEL E. SENKO
Manager, Information Sciences
International Business Machines Corporation
San Jose, California

HON. ROBERT C. WOOD
Under Secretary
Department of Housing and Urban Development
Washington, D.C.

DR. EMANUEL S. SAVAS
Deputy Chief Administrator
City of New York

DR. EDWARD H. ERATH
Technical Advisor Mayor's Office *and*
President
Los Angeles Technical Services Corporation
Los Angeles, California

DR. CHARLES DRESCHER
Manager, Information Systems
Los Angeles Technical Services Corporation
Los Angeles, California

DR. NICHOLAS E. GOLOVIN
Office of Science and Technology
Executive Office of the President
Washington, D.C.

MR. JAY W. RABB
Deputy Director
National Airspace System Program Office
Federal Aviation Administration
U.S. Department of Transportation
Washington, D.C.

MR. F. STEWART BROWN
Chief, Bureau of Power
Federal Power Commission
Washington, D.C.

DR. JAMES M. PITTS, JR.
Department of Chemistry
University of California
Riverside, California *and*
President
Environmental Resources, Inc.

DR. STANLEY M. GREENFIELD
Head, Environmental Sciences
RAND Corporation
Santa Monica, California

Introduction

A major reason for the existence of the American Society of Cybernetics is the need to bring together scientists from many disciplines for the purpose of working on problems which cannot be solved within a single field of research. In particular, these problems include those which are of the greatest concern to our future: the population explosion, pollution, the stability of economic systems and government, the problems of war and violence, and, finally, the complex problems associated with large-scale transportation, communication, and power distribution systems. The common denominator of these problems is the fact that they all involve large-scale systems.

The conference described in the following pages represents one of the first interdisciplinary attempts along these lines. The papers fell naturally into three categories. The first category concerns discussions of a somewhat philosophic nature that indicate general directions to solutions of certain problems and point out some problems that have been overlooked in the past. The second category deals with the future of large-scale computer systems. These papers consider both hardware and software developments, as well as social implications. The third category of papers, which, incidentally, is the largest category in this volume, involves the direct application of cybernetic or systems techniques to such important real problems (not merely theoretical exercises) as: government, transportation, power distribution, and control of weather and pollution.

Before summarizing these, however, I must point out that certain of the burdens and responsibilities usually placed on the shoulders of an editor were absent in my case. I did not pick the speakers for this conference. This was done entirely by Dr. Lawrence J. Fogel (President of the ASC). Secondly, it was decided by the Board of Directors of the ASC that all the talks would be published, i.e., that there would be no subsequent selection of papers from those given at the conference. My first duty was, therefore, to request manuscripts from the speakers. To my amazement, every speaker sent in a finished manuscript before the final deadline—there were no exceptions. In all cases their manuscripts accurately represented the material given at the conference, and in several instances they were written with the help of the typed transcriptions from tape recordings.

The first four papers of this conference contained discussions of certain general philosophical issues. Professor Harold Hughes discussed utopias in history and showed how they served the purpose of setting goals for the society of the times in which they were created. He also pointed out their important shortcomings in the sense that they probably could never be realized in practice. In contrast, he described the necessary properties for planned societies, based on principles of systems analysis.

The paper by Professor Donald Walter concerned itself mainly with the problem of free will and cybernetics. First, he expressed concern over the application of cybernetics to social systems: there is danger in a comprehensive collection of data on all individuals in a social system. Such a thing might be too great a temptation to present to people in power and would be a step toward totalitarianism. In addition, Dr. Walter feels that, in general, it might not be appropriate to make approximate engineering models of social systems. The main difficulty he attributes to the problem of modeling "free will" on a computer. He suggests that the problem should be treated stochastically, and he draws a very interesting analogy between the social problem and the problem of understanding brain waves in connection with molecular memory (a problem which is fascinating in itself).

Professor Stephen Sherwood discussed the problem of defining the word "normal" in medicine in relation to society. This, of course, will be necessary if one is to perform medical diagnosis with the aid of a computer. The main point of his paper is that the concept of "normal" can be defined only in the context of the goal of the society within which the patient lives. He related the concept of "abnormal" to an "error" that needs to be corrected in the context of feedback control. (In a similar connection, the reader may also wish to read a paper entitled "Tragedy of the Commons" by G. Hardin in *Science 162*, 1243, 1969.)

In the last paper of this section, Professor Thomas Naylor presented a research proposal for the computer simulation of a number of different types of models to evaluate alternative population planning policies. This paper implies that merely instructing populations to cut down the size of their families is incredibly simplistic. The problem is complex; it is felt the creation of involved mathematical models might lead to intelligent decisions regarding policy. The 57 references given at the end of this paper should provide an excellent bibliography for anyone interested in this problem.

The section on large-scale computers commenced with the paper by Professor John Kemeny who emphasized the need to avoid certain pitfalls in the use of the computer for simulation of social systems. One of these pitfalls is the belief that, if a model is not in closed analytic form, it is not a proper mathematical model. Professor Kemeny explained, on the basis of his own research experience, that such thinking can be very misleading. He suggested that one should, in fact, investigate algorithmic or "computer code" types of models. This, he implied would be the most feasible method until the time when the field of mathematics is developed to the point where it is sufficiently complex to deal with such problems. He made several suggestions regarding the use of computer models. These seemed mainly concerned with the necessity of continuous on-line interaction between the modeler and the computer. He gave illustrations of what misfortune can occur when certain general principles of such on-line interaction (i.e., feedback) are ignored. Finally, he proposed that when computers are employed for the purpose of supplying people with information, the computer should serve a dual purpose. More specifically, as it gives information, it should also obtain data (from the population) which can then be subsequently utilized by the model-makers.

The next three speakers (who work in the computer manufacturing industry) gave a preview of the computers of the next generation. Their papers had certain points in common. They were all imaginative and optimistic while remaining realistic. On the basis of trends and present plans, they give projections or predictions of what we should see in the '70's, regarding new hardware and software. In general, they say we should expect a tremendous increase in the availability of on-line remote terminals to computers which would permit complex forms of interaction between the terminals of the system. They also predicted a much larger flexibility and "ability to adapt" in the executive programs and in the "languages" used to program a computer. Mr. Earl Joseph pointed out the possible use of an analogy from biology which involves DNA and RNA code transfer processes and indicated that an analogous process might be useful in large-scale integration. He also predicted that, in the future, there would be "paperless books" (i.e. books in the form of computer output). In fact, as Charles Purcell pointed out, centralized communal computers would supply information not only from books, but from news media and other information over a wide range which would be put into the form of printed paper at the terminal. Drs. Michael Senko and Arthur Anderson, both of IBM, pointed out the special importance of developing languages that will allow the user to tell the computer *what* he wants to be done, while the program itself would take over the job of determining *how* the computer should do it. They also gave examples of how the computer can interact between individuals and their organizations and showed how large corporations could benefit from such interactions. Similar ideas were also discussed by Mr. Joseph.

If the predictions of these gentlemen are even approximately correct, it appears that, in the '70's, the control and dissemination of many types of crucial information will involve computers to a vast extent. It seems to me that the transition will have to be made very cautiously and that certain precautions would be necessary to prevent serious crises. Besides the need to avoid incredible "goofs" that can be made by the computer (as can be seen from the accounts in the newspapers), there is also the need to build in relatively complicated safeguards when, for example, computers are used very extensively in the banks. Embezzlement by computer has already been accomplished. It would thus appear that it is possibly a mistake to train prisoners in the subject of computer programming (as is currently being done). It is clear that when we apply large-scale computers to social needs we will suffer very greatly if we have to learn from our mistakes. Rather, it is necessary that we anticipate ahead of time, as much as possible, the things that can go wrong through programming error or through criminal behavior. Of course, this sort of observation was made as long ago as the Forties by Norbert Wiener.

The third category of papers dealt directly with actual problems now faced in the present. They all created a special interest stemming from their direct contact with reality. The first three came under the general heading of Governmental Systems. A paper by the Honorable Robert Wood was concerned with the urban problems. It was given as an after-dinner talk following the banquet. The emphasized point was that science is obviously needed to cope with the gigantic problem of the cities; however, the real problem is that the public regards science in a bad light and is very skeptical that science can be of any help. Therefore, it is difficult to convince Congress and the taxpayer that science can indeed play a helpful role. Briefly, it is not a question of the scientist turning from whatever he is doing at present to apply himself to social problems but, in fact, his first convincing the public that he can be useful.

The paper by Dr. Emanuel S. Savas contains several surprises. First, he described the government of New York City in terms of a control system block diagram. This set the tone for the rest of his talk which was entirely in terms of feedback, control, and so on. One could call his talk either a general systems approach, a cybernetics approach, or an operations research approach. The most surprising thing that emerged from this confrontation with New York City's urban problems, using a scientific approach, was the relative ease of finding solutions to problems, in comparison to the tremendous difficulty of putting these solutions into effect. It seems that one of the most important things to accomplish, if we are to solve the crucial urban problems, is the setting up of governmental machinery that will allow one to put into effect the solutions to the problems. Dr. Savas's supposedly apocryphal story of a boiler and a building inspector illustrated the important point that when one deals with real social systems, the problems that one faces are sometimes totally unexpected and involve things which would probably never be anticipated. The paper left a strong impression that this field is a fascinating one, if only for the surprises it affords. Dr. Savas's paper gives many illustrations of how one can use feedback concepts to alter governments and make them more effective. It seemed to give a realistic indication that cybernetics concepts can be used in a very practical way for such problems. I understand that the Russians are doing similar things, perhaps on a wider scale, in their government.

The next paper, by Dr. Edward Erath and Charles Drescher, also concerned urban problems from a systems engineering point of view. These authors point out that it's more complicated to design an effective transportation system for our country than it is to put a man on the moon. Like Dr. Savas, they recommend the use of cybernetic or systems approach to solve these problems. They point out the analogy between the military establishment and the typical police department and show the contrast between them, the military being systems-oriented and the police departments being relatively obsolete in their approach. These authors agree with Dr. Savas that the present situation in the cities is appalling in many respects and that we already have the scientific knowledge to make tremendous improvements.

The next paper in the section on governmental systems was given by Dr. Nicholas Golovin and is here published posthumously. It represents the culmination of 15 years of work and the final contribution of a brilliant individual. His main point was that our nation has become so complex that our government cannot presently cope with information needed to plan its activities. He therefore proposed a "fourth branch of government" to perform the functions of data collection and evaluation necessary for the other three branches. He emphasized that a number of precautions must be taken to prevent special interests from getting control of this fourth branch. In private conversations with him I learned that he had much more to say than he wrote here concerning the methods of maintaining the important "checks and balances" structure (feedback) between the proposed branch and the present three. The premature death of our colleague is a great loss to our nation and to our intellectual community.

The last four papers in this volume dealt with the problems of transportation, electric power, and environmental control. Jay Rabb described both the present system for air traffic control and the one planned for the future. The FAA intends to switch from what is predominantly a "manual" system to a system which is almost entirely automatic and which involves computers. The system of the future will still involve a large number of people, however, and one of the great difficulties in preparing the future system is, again, the fact that people are unpredictable and one cannot simulate them adequately in a mathematical model. At one point Mr. Rabb expressed his sympathy to those who will have to simulate the human part of the systems of the future. Later he gave an illustration of how incorrect a computer simulation of human behavior can be. The anecdote does not appear in the written version of his talk. Sometime in the early '60's, a projection was made (for 1970) of the number of instrumented landings and take-offs to be expected at airports. This prediction came true on the following year. What had been left out of the model was (1) the increased size of aircraft when the change was made from propeller to jet planes and (2) the sudden collapse of the competition from the rail system. These two factors gave a sudden and large change in the number of flights. After hearing this illustration, I began to appreciate in a more scientific way the problems that face those parts of the government directly and indirectly involved with trying to predict the needs of the population. This unpredictability of population behavior seemed to be one of the emergent themes of the conference.

The paper by F. Stuart Brown was especially interesting in view of the famous power failure in the Northeastern United States in November. 1965. At one point in his paper Mr. Brown explained the cause of this power outage. In essence, it seems that the threshold setting of one relay was not at the proper value and that this started the entire avalanche.

This single event was simply unforeseen by the designers of the system. It certainly illustrated the vulnerability of large-scale systems and the instabilities that can lurk unseen until an unexpected moment. Mr. Brown explained many of the precautions that are now already in effect, but he also pointed out the still-unsolved problems, such as the nature of the load which can be predominantly inductive, capacitive, or resistive, depending on what sort of appliances are turned on at any particular time of the day. The final upshot of the paper is that the reliability of power systems is a very complex and challenging and still-unsolved problem.

It should be added that this particular problem has an added fascination to it due to Norbert Wiener's use of the analogy between the large number of interconnected alternating-current generators and the oscillators responsible for the electroencephalogram from the human brain. Though this analogy has never been proven to be correct, it is certainly correct for other biological rhythms, such as the so-called circadian rhythms (those with natural rhythms of about 24 hours). A long jet trip in the easterly or westerly direction into other time zones can have the effect of causing certain oscillations in the body to become temporarily mutually desynchronized in a way that can resemble a power system "going unstable." Information about the stability of power systems could therefore have potential value in biology (c.f. the behavior of communities of fireflies, or heart cells).

The last two papers had to do with our environment. The first of these was by Professor James M. Pitts, Jr. This talk is concerned mainly with the problem of air pollution, and, in particular, the effects of automobile exhaust fumes on the atmosphere in the vicinity of Los Angeles. There are several different types of smog and the smog caused by automobiles has a number of chemicals in it which are particularly noxious. By showing the chemical constituents of the smog as a function of the time of day, Dr. Pitts described a theory that would explain the formation of certain chemicals as being due to photo chemistry—that is, the effect of the sun on the raw exhaust fumes that rise into the atmospheres from the Los Angeles freeways. Dr. Pitts also explained some laboratory experiments that simulated these effects and he thus explained to some extent the existence of certain of the gases in the smog, especially at the end of a day. He likened this pollution to a form of gas "attack." All present were struck with the challenge that the pollution problem gives to scientists who want to become socially involved.

The final paper in these proceedings was by Dr. Stanley Greenfield. He concerned himself primarily with weather modification and discussed for example, the use of seeding to influence clouds and precipitation. Dr. Greenfield pointed out that while modeling of weather systems and field experiments have both been carried out at this time there seems to be a remarkable lack of intercommunication between these two approaches and that much more communication must take place if there is to be any successful weather modification in the future. The tone of this paper was somewhat pessimistic, although I realize from personal conversations that the author is perhaps secretly optimistic. Dr. Greenfield seemed to be concerned primarily with two aspects of the problem. The first was with the previously mentioned collaboration between the theoreticians and the experimentalists. His second and greatest concern was the enormous danger that is involved in the future of weather modification. He perceives the possibility that man, in his ignorance, might destroy his own planet by making a small mistake which becomes greatly amplified. In essence he perceives that we are in the position to cause, through ignorance of the highly complex and nonlinear system that is responsible for the global weather, an irreversible transition to an Ice Age. In addition to these two major points, Greenfield explained some of the interesting details of the mechanism of the global weather system, especially those relating to energy transfer.

Although a large number of disciplines were represented in this conference, one clear message emerged. This was partially expressed in varying forms in most of the papers. Civilization is rapidly becoming very dependent on large-scale systems of men, machines, and environment. Because such systems are often unpredictable we must rapidly develop more sophisticated understanding of them to prevent serious consequences. *Very often the ability of the system to carry out its function (or alternatively, its catastrophically failing to function) is a property of the system as a whole and not of any particular component.* The single most important rule in the management of large-scale systems is that one must account for the entire system; the sum of all the parts. This most likely will involve the discipline of "differential games."

It is reasonable to predict that cybernetic methods will be relevant in the future for the solution of the greatest problems that face man today. It is difficult to image how such complex problems will be solved without an unprecedented amount of collaboration between the different disciplines which will allow different scientists to work together, utilizing each other's expert knowledge and methods. There is much to be done.

EDMOND M. DEWAN, PH.D.
Bedford, Mass.

I
General Aspects

Utopias and Cybernetic Cultures

HAROLD K. HUGHES

State University of New York
Potsdam, New York

I. Introduction

In a time when institutions are being challenged more widely than in any period of history, including the French and Russian revolutions, it is wise to step aside from the daily confusion to view our large social organizations from the perspective of history.

In this article, I am examining utopias as one form of social protest in fact and fiction for the light they may shed on our own age, hoping to sort out what can be from what utopians dream should be.

This examination is undertaken in the same spirit which motivated the National Planning Association to study the economic feasibility of achieving the utopian "Goals for Americans" which were set forth by the Eisenhower Commission in 1960. The Association found that the cost of national objectives generally backed by most Americans would exceed the nation's resources by $150 billion annually in 1975. Therefore, all of these goals cannot be reached by that year; this would be true even without the Vietnamese war.

It is particularly appropriate to address these remarks to the American Society for Cybernetics, the first group of professional utopians in history to start down the long road to a cybernetic culture.*

As an activist, as well as an amateur social philosopher, I have tried to achieve a measure of objectivity by resort to history, which is to sociology what laboratory experimentation is to science. I can hope to be successful only in part for, as I read history, it is impressed upon me that each utopian seeks to cure only the worst outrages of his own culture. On the other hand, my hope to be objective is reinforced by the certainty that human nature has not changed over recorded history. There now exists a vast record of man's follies, successes, loves, and brutalities which can be the raw data for this study.

The principal task of the scientist is to deal with the unknown. Now that man can control so much of nature, he must learn about the nature of man, that greatest reservoir of unknowns, so that he can control himself.

Omar Khayyam, in the *Rubaiyat,* tells us of his own search for a philosophical utopia:

> Myself when young did eagerly frequent
> Doctor and Saint, and heard great argument
>   About it and about: but evermore
> Came out by the same door wherein I went.

I trust that this lovely quatrain does not make you think we are setting off in search of unintelligible answers to unasked questions.

II. Utopian Cultures

Definition of Utopia

For the purposes of this article, I define a utopia as a culture, supposed by the author to be more desirable than the one in which he lives, incorporating all of his goals as ultimate and generally static values. I shall sharpen this definition later.

---

*Editor's note: The editor does not share the author's optimism. Avoidance of disaster seems to be our present goal.

"Without the Utopias of other times," said Anatole France, "men would still live in caves, miserable and naked. . . . Out of generous dreams come beneficial realities. Utopia is the principle of all progress."

Frederick Engels made an interesting distinction between utopian and scientific socialism, which illuminates the political thinking characterizing communism from the 1840's until Russia began to supplement Marxism with cybernetics after Stalin's death. A utopia, according to Engels, is any social scheme which does not recognize the division of society into classes, the inevitability of class struggle, and social revolution. By this definition the United States now is a utopia.

No Women Utopians

Throughout recorded history men have created hundreds of utopias, most of them imaginary. I have yet to run across a woman utopian, although many, like Margaret Mead and Marie Louise Berneti, have written about them.

Bulwer Lytton remedies this oversight by making women stronger and more skilled in the use of the ultimate weapon, vril, in his 1870 novel, *The Coming Race*.

Utopias Are Generally Isolated

Most utopias, like the locales of James M. Barrie's plays, are purposely isolated from the rest of humanity. In about 850 B.C. Lycurgus, the "Lawgiver of Sparta," took drastic steps to reduce commerce and communication with other Greek cities. For example, he replaced gold and silver coins with iron ones of low value which were accepted in Sparta but ridiculed in the rest of Greece.

Citizens could not leave More's 1516 Utopia any more than they could leave Stalin's Russia. Thomas Aquinas, advocating self-sufficiency for his city, observed in 1260, "if the citizens themselves devote their lives to matters of trade, the way will be opened to many vices."

Johann Valentin Andreae placed his 1619 Christianopolis

on an island with strict immigration controls. Denis Diderot's 1772 utopia was on the real, but then largely unknown, island of Tahiti.

The Size of Utopias

In keeping with their isolated location, most fictional and all real utopias have had relatively small populations controlled by a single individual or a small oligarchy. Andreae's *Christianopolis* had a mere 400 inhabitants. Notable exceptions were More's Utopia whose 5.4 million population duplicated England's of his day, and Gabriel de Foigny's 144 million on Australia in 1676. Foigny apparently was unaware that the entire population of the world was then only 500 million. Australia's present population is but twelve million.

In his 1948 best seller, B. F. Skinner populated Walden Two with a mere 1000, although Walden Six was being colonized and plans for Walden N were in the making.

Utopian Sex

For all the variety of emphasis in utopias, sex is one subject that almost always receives extensive attention. From the continence which killed off the Shakers, through W. H. Hudson's sexless "Crystal Age," to Campanella's free love for men before age twenty-one, just about every conceivable norm has been advocated by some author.

There were eugenic marriages in Plato's imaginary Republic and in real Sparta:

> If Lycurgus discouraged love, as we understand it, between men and women he did, on the other hand, take great pains to see that women should be sexually attractive to men, and instituted public dances and other exercises of naked young maidens in the presence of the young men, not only to take away the excessive tenderness and delicacy of the sex, the consequence of a recluse life, but also because these exercises were incentive to marriage.[1]

Both the real Spartans and the imaginary Republicans practiced infanticide if, after examination by the most ancient men of the tribe, a child was found to be weak or deformed.

More's Utopians had an interesting engagement custom. "Before marriage some grave matron presents the bride naked whether she is a virgin or a widow, to the bridegroom; and after that some grave man presents the bridegroom naked to the bride. We indeed both laughed at this, and condemned it as very indecent. But they, on the other hand, wondered at the folly of the men of all other nations, who, if they are but to buy a horse of a small value, are so cautious that they will see every part of him, and take off both his saddle and all his other tackle, that there may be no secret ulcer hid under any of them; and that yet in the choice of a wife, on which depends the happiness or unhappiness of the rest of his life, a man should venture upon trust, and only see about a hand's-breadth of the face, all the rest of the body being covered, under which there may lie hid what may be contagious, as well as loathsome."

Possibly because of his fear of the Geneva Venerable Company (censors) or to twit them slyly, Foigny had his Australians go about perpetually naked, but then he made them all of one sex, hermaphrodites.

We see, then, that public sexual stimulation is ancient and as universal as we might have supposed from today's miniskirts and Margaret Mead's description of the Manus' phallic shell dance.

Utopian Freedom vs. Conformity

If ancient Greece, instead of being a loose federation of free cities, had been a strongly centralized country, such as Plato advocated in his *Republic,* then Homer, Sophocles, and other great writers would not have been able to produce their masterpieces. Yet, because Greece was inadequately organized to resist its neighbors, it lost its freedom and never regained it, even to this day.

With the notable exception of Francois Rabelais' *The Abbey of Theleme* (1535), there is a curious mixture of extreme rigidity and license in most utopias, both real and literary. That great advocate of progress, Francis Bacon, maintained his *New Atlantis* (1627) with secret police and an elaborate spy system. Since man is not by nature a social being with moral sense, in Thomas Hobbes' *Leviathan* (1651) the state has a right to absolute power. Even that great humanitarian social reformer, Gerrard Winstanley, who wrote *The New Law of Righteousness* in 1649, advocated very rigid laws and prescribed slavery as punishment for minor crimes.

In general, then, we see utopias correcting specific evils but failing to recognize man's march toward democracy in all of life.

Attitude Toward Work

Along with sex, all utopias specify rules for work except for the favored dilettantes in Rabelais' Abbey. Inhabitants of Tommaso Campanella's *The City of the Sun* (1602) worked a mere four hours per day, while Etienne Cabet's *Iscarians* raised this to five. It seems to me that Bulwer Lytton's solution to the work problem in *The Coming Race* (1870) is the most vicious; children do all the tedious, unpleasant, and dangerous work. This reflects the last stages of England's industrial revolution at its worst.

The shortage of work in 1969, particularly for the unskilled, continues to plague developed countries. We have made no real advance in the happy use of leisure time such as that achieved by the fictional characters in Skinner's *Walden Two* (1948).

The Changing Utopian Image

From the days of Lycurgus (about 850 B.C.) utopian thought shows a steady progression, matching the social images of its creators. At first they focused on health and a plenty of goods when subsistence was marginal. Then, as agriculture improved, personal and religious freedom became the cherished goal.

Can we say today that because of overpopulation and our anonymous dependence on machines we seek utopias to give us life's meaning in some holistic sense?

Tommaso Campanella was poor and spent much of his life behind Inquisition bars. Thus his *City of the Sun* has no torture, no prison, and equality of goods for all. He would, however, condemn to death any woman who used makeup or high heels.

Francis Bacon, in contrast, was rich and sought political power. Hence his *New Atlantis* kept private property, sumptuous living, and the monarchy. Only a scientific research institute was added, challenging no established interest.

Living under the excessive repressions just preceding the French Revolution, Diderot even renounced the right to decree that "each should do what he wills," which is the ultimate Goedelian paradox of liberty.*

Capek's *RUR* (1928), Aldous Huxley's *Brave New World* (1932), and George Orwell's *1984* (1948) are examples of recent anti-utopias. By their very negativeness and popularity they tell us something about our quest for meaning. None is concerned with food, shelter, or clothing.

III. Cybernetic Cultures

Space does not allow a further description of utopian cultures, so I now proceed to contrast these with cybernetic cultures.

---

*Editor's note: The author is making a partial analogy here with two remarkable theorems due to Kurt Gödel. The first or "Unprovability" Theorem says that if a formal system is "consistent" in a certain sense, then there are undecidable but nevertheless "true" formulae in the formal system (i.e., formulae which cannot be proven to be true or false within the system). The second theorem proves the undecidability within the formal system of a formula expressing the "consistency" of the system. In other words, the consistency of a system cannot be proven starting only with its "axioms" and using only its "rules of inference." Further information will be found in *Gödel's Proof*, E. Nagel and P. Newman, (New York University Press, 1958); and *On Formally Undecidable Propositions*, Kurt Gödel translated by B. Meltzer (Basic Books, Inc., (1962).

Definitions*

1. A *system* is a collection of interacting, diverse elements which function (communicate) within a specified environment to process information in order to achieve one or more desired objectives. Feedback is essential. Some of its inputs may be stochastic and a part of its environment may be competitive.

2. The *environment* is the set of variables which affects the system but which is not controlled by it.

3. A *complex system* has five or more internal and nonlinear feedback loops.

4. In a *dynamic system* the variables or their interactions are functions of time.

5. An *adaptive system* continues to achieve its objectives in the face of a changing environment or deterioration in the performance of some of its elements.

6. The rules of behavior of a *self-organizing system* are determined internally but modified by environmental inputs.

7. *Dynamic stability* means that all time derivatives are controlled.

8. A *cybernetic system* is complex, dynamic, and adaptive. Compromise (optimal) control achieves dynamic stability.

9. A *real culture* is a complex, dynamic, adaptive, self-

*****Editor's note:** Some of these concepts (system, dynamic system, adaptive system) can be defined in alternate ways. Some of the most general and mathematically precise definitions of "dynamical system," for example, will be found in *Topics in Mathematical System Theory* by R. E. Kalman, P. L. Falb, and M. A. Arbib (McGraw-Hill, 1969, p. 5), or *Qualitative Theory of Differential Equations* by V. V. Nemytskii and V. V. Stepanov (Princeton, 1960, p. 328). A tremendous literature exists in the area of systems (all potentially useful for understanding living systems) and range from introductory books such as *Signals and Systems in Electrical Engineering*, W. A. Lynch and J. G. Truxol (McGraw-Hill, 1962) and *Control System Synthesis*, J. Truxal (McGraw-Hill, 1955) for the classical information to more advanced modern books as *Modern Control Theory* by J. T. Tou (McGraw-Hill, 1964), *Modern Control Principles and Applications* by J. Hsu and A. Meyer (McGraw-Hill, 1968) and *Optimization, Estimation and Control* by A. Bryson and Y. Ho. (Blaisdell, 1969).

organizing system with human elements and compromise control. Man is in the feedback loop.

10. A *cybernetic culture* is a cybernetic system with internal rules, human elements, man in the feedback loop, and varying, competing values.

11. A *utopia* is a system with human elements and man in the feedback loop.

These definitions are summarized in Table 1 where 1 means always present, and 0 means generally absent.

Table 1

Characteristics	System	Complex System	Dynamic System	Adaptive System	Self-organizing System	Cybernetic System	Real Culture	Cybernetic Culture	Utopia
Collection of interacting, diverse elements, process information, specified environment, goals, feedback	1	1	1	1	1	1	1	1	1
At least 5 internal and non-linear feedback loops		1				1	1	1	0
Variables and interactions functions of time			1			1	1	1	0
Changing environment, deteriorating elements				1		1	1	1	0
Internal rules					1	0	1	1	0
Compromise (optimal) control						1	1	1	0
Human elements						0	1	1	1
Dynamic stability						1	0	1	0
Man in feedback loop						1	1	1	1
Values varying in time and competing						1	0	1	0

Comparison of Utopias and Cybernetic Cultures

Table 2 contrasts some of the characteristics of utopias and cybernetic cultures.

Table 2

Characteristic	Utopia	Cybernetic Culture
Size	Small	Large
Complex	No	Yes
Environment	Static, imaginary	Changing, real
Elements deteriorate	No	Yes
Rules of behavior	External	Internal
Control	Suboptimized	Compromised
Stability	Static	Dynamic
Values	Fixed	Varying, competing
Experimentation	None	Evolutionary operation

IV. Social Relativity

Over the past million years of man's cultural evolution he has constructed many different yet stable societies. There is, thus, a relativity principle in sociology with a meaning similar to its meaning in physics: Value judgments are arbitrary.

Given a set of values, in a given physical environment, and given specified competitors for the same resources, the culture is uniquely established. Change any one of these boundary conditions (constraints) and you change the culture. Of these constraints the values seem to be most flexible in number and intensity, the culture adjusting them, Parkinson-like, to reduce the residual degrees of freedom to near zero.

It is helpful to think of the determinants of any real culture as you would the solution of a set of independent simultaneous equations in many unknowns. Society keeps adding constraints (equations) until their number equals the degrees of freedom (the number of unknowns) in the system. Then the culture is determined.

It is a purpose of this article to identify a small number of necessary but not sufficient constraints which seem to be present in modern cultures, yet frequently are overlooked. Hopefully, these will guide political activists as they advocate their programs.

The propositions I am going to present may be restricted by the boundary conditions of the United States in 1969. Some of these conditions are easy to identify while others, maybe a majority, are buried in our unconscious like the nine-tenths of the proverbial iceberg beneath the surface of the ocean.

Other human ecologies, possessing different boundary conditions, arrive at other social structures. Yet, there is reason to suspect that the diverse forces which are changing Nasser's Egypt, Mao Tse-Tung's China, and Richard M. Nixon's America are pointing toward similar cybernetic cultures and it is interesting to speculate whether we shall, in time, all arrive at a homogeneous world.

V. Restraints on Cybernetic Cultures

Man Constructs Utopias

My zeroth constraint or boundary condition is that man is committed to construct utopias. From the days of Sparta through Rapp's New Harmony to the promised lands of last fall's campaign, man has created magnificent fictions of perfection and tragic realities of imperfection. Until our age, these were but perturbations on an unheeding ecology. Now, however, his aggressions and his pollutions are first-order threats to his continued existence. In desperation, he is turning to science, hoping to discover a new road to the old destination.

> We are such stuff
> As dreams are made on...

said Shakespeare.

By the standards of the underdeveloped and non-egalitarian American colonies of the eighteenth century, the Great Society in which we live has already reached their goals.

In tackling the complex field of social ideals you may observe that I share Oscar Wilde's view: "A map of the world that does not include Utopia is not worth glancing at. . . . Progress is the realization of Utopias."

Diversity

My first constraint derives from the observation that there is a great variability in all living structures and social organizations.

There are roughly 5000 different kinds of trees and no one dares to enumerate the varieties of insects. Possibly they exceed two million. One might think that within any one species the members would be identical, but this is far from true. Let our own bodies be an illustration.

There are eight blood groups, some of which may not be transfused into the others. Our taste buds respond differently to creatine and other chemicals. Our stomachs vary in size by a factor of fifteen, our hearts by a factor of two. Normal pulses range from 45 to 105 beats per minute, and our blood flow is considered normal between the limits of three and eleven liters per minute.

Eight different patterns of tendons on the backs of our hands are known, and thyroid glands may weigh anywhere between eight and fifty grams. Even the number of parathyroid glands is not the same but varies from two to twelve, each weighing from 50 to 300 milligrams. Our daily food requirements likewise vary widely, calcium by a full 500 per cent, for example.

We know, of course, that IQ varies on either side of the population standard of 100 with a sigma of fifteen. On this scale individual IQ's range from zero for the non-trainable, vegetable types at Muscatatuck State School in Indiana, to geniuses with IQ's in excess of 160.

It is scarcely a wonder, then, that signatures differ, that no two people think exactly alike, and that one man's play is another man's work. (Or should I say, One man's secretary is another man's wife?) It may even be that our vote for Humphrey, Nixon, or Wallace was a consequence of our anatomy and physiology as much as it was the dollars that each spent in the campaign.

Why do some persons become alcoholics? An accumulation of evidence throughout centuries shows that this is a disease which attacks certain individuals and capriciously misses others who are evidently invulnerable. A careful study of 1000 cases has shown that 105 of them were intoxicated (as determined by objective tests) when the blood alcohol level was 0.05 per cent, and at the other extreme, sixty-seven individuals were sober when the alcohol level in the blood was eight times higher, 0.40 per cent.

It is not surprising that our bodies, like our automobiles, have defects. Rather, the wonder is that they work as well as they do, considering their great complexity and variety. But nature has had two billion years to weed out its mistakes. No man-made device has such a high reliability nor an equal capacity for self-repair.

I would like to extend this constraint beyond the observation that all biological and social structures are variable and hypothesize that all viable societies must permit and even encourage diversity. Utopias allow no diversity and, therefore, cannot survive. In the words of Robert Browning, "What comes to perfection perishes."

As institutions are born, so they must die, particularly in the static-population society ahead after about the year 2000. Since neither their birth nor death is ever sudden, there is a period of coexistence and, therefore, a continuous diversity.

As illustrations of this constraint, I conclude that a single world-wide religious organization is not desirable. Indeed, I hypothesize that even a single Christian church is theoretically impossible, unless its structure be so loose and its elements so diverse that they strain the meaning of the word ecumenical.

In every group, whether it be a university, the American Society for Cybernetics, or the National Association of Social Workers, we see or shall see splintering. I hypothesize that this splintering enhances the viability of the group and is inevitable.

By entirely empirical procedures, the American people have come to the same conclusion with respect to state and local government. In spite of the predominant attention paid to the growth of the federal government, it is a fact that the ratio of state plus local taxes to federal taxes has been climbing steadily. This ratio is now about one and the relative effect of the federal budget on the economy, in comparison with the effect of local and state budgets, is less today than at any time in the past.

Other statements of this variability constraint may make its meaning clearer. As no vaccine protects against all disease, and as no football play always produces a touchdown, so there is no single solution to prostitution nor to any other major social problem. We say in science, "avoid the monistic error." Should I add the "monastic error" also?

Our colleagues in the life sciences teach two other forms of the diversity constraint under the names "the competitive exclusion principle" and "the axiom of inequality." These deny that complete competition can exist or that any two living organisms or processes are precisely equal.

In a similar vein, Hutchinson wrote about a small fresh water crustacean called Daphnia: "organisms actually do need randomness in the environment and can't get on properly without it." [2]

The Biblical injunction to produce equality, "Sell all that you have and distribute to the poor," is an obvious paradox, impossible for all to obey.

I would like to quote from a very interesting article by Ronald L. Lee called "The Paradox of Equality: A Treat to Individual and System Functioning," wherein the author says:

> an egalitarian model of society is not a suitable or efficient idealogical model for a rapidly industrializing society . . . it is approached only in small social groups where economic production leaves little or no surplus beyond that essential for the maintenance of its members . . . it dictates the reduction of social span and is thus in contradiction to the principles of system growth . . . if a cultural system does not support the division of labor . . . strains may be built into the society.[3]

In contrast, most utopias specify equality of income, but these were written in periods of scarcity.

Some years ago Pareto advanced his law of social stability. In effect he gave upper and lower bounds to the standard deviation of family income within which a culture would be stable and outside of which it would disintegrate.

The utopian spirit of equality was present at the writing of the U.S. Constitution in Philadelphia: "We hold these truths to be self-evident, that all men are created equal. . . ." I remember my schoolboy distress on the first occasion that I failed to defend this shibboleth.

It is important to observe that diversity means a spread around some central value, not a random aggregate of widely divergent values. Members of a coherent society must have some common beliefs as well as a tolerance for subordinate beliefs of others which are at variance with their own subordinate beliefs. I interpret much of the recent student agitation in colleges as a challenge to that statement, a challenge that gives society no alternative but to force adherence to lawful procedures for change.

A part of the diversity in a cybernetic culture must be stochastic or wasteful in order that there be sufficient deviation from the goal to bring error control into play.

The diversity constraint on cybernetic cultures is in sharp contrast to what seems to hold for inanimate matter. The universe appears to be homogeneous in physical laws. We speculate that this is true throughout both space and time. For example, the fundamental physical constants are the same today as they were ten billion years ago when the "big bang" started the expansion of the universe, and they are the same on earth as in the most distant galaxy. All electrons are identical in every respect, but no two persons in all of history have ever been the same.

To close this section, I would like to observe that there is also an upper bound on diversity, as there is on freedom. Each new generation tests this bound and each age sets its limits on conformity. We have only to reflect on the numerous confrontations between students and college administrations, and the violent black protests to understand why "law and order" and George Wallace attracted so much support in the recent election. Said Thomas Huxley: "Life cannot exist without a certain conformity to the surrounding universe." That old conforming radical, Benjamin Franklin, had this advice to give us: "Singularity in the right hath ruined many; happy those who are convinced of the general opinion."

Original Ignorance

The conscious mind of man is a clean slate at birth, but many of his instincts are well-developed and appropriate to his jungle ancestry. Thus his survival is a race between education and catastrophe.

As the second constraint on possible cybernetic cultures I take this statement:

> We hold these truths to be self-evident, that all men are created ignorant, that they are endowed by their creators with certain inherent variabilities, that among these are health, adaptability, and the pursuit of power.

Alternately, we may end this statement: that they are endowed by their creators with certain inalienable drives, that among these are wife, bonhomie, and the happiness of pursuit.

To focus on the fact that all men are born ignorant seems trivial until we recognize that its logical consequence, often overlooked by reformers, is that much of the wisdom in society resides in its institutions. Every cybernetic culture must, then, have a time delay built into its formula for modification. It is not always wise to change courses in the middle of a dream, and we must never permit instant, electronic voting on public issues when they arise.

You probably have heard that when worms and rats which are conditioned to certain behavior patterns are fed to other worms or rats, the fed animals acquire those same behavior patterns.* It seems unlikely, however, that at any time in the

---

*Editor's note: This is still an extremely controversial subject and it appears that most psychologists and biologists do not believe that such a transfer of memory has been convincingly demonstrated at this time.

next century we humans shall progress to such cannibalistic learning, even if my friend, Dr. Lawrence Fogel, does will his brain to the next generation of students. There is not even any hope that General Foods will be selling Physics 205 pills.

So great a laziness attends our ignorance that past and present societies have found it necessary to apply the full force of their authority to requiring people to learn. As Arthur Schlesinger observes:

> Our perceptions of reality are crystallized in a collection of stereotypes; and people become so fond of the stereotypes, so much at home with them, that they stop looking at actuality. In this way they protect themselves from the most painful of human necessities, which is, of course, the taking of thought.[4]

The Irrationality Constraint

According to Mark Twain, "man was made at the end of the week's work when God was tired." This may explain why man "is a rational animal who loses his temper when he is called upon to act in accordance with the dictates of reason." [9]

There is no doubt that, in contrast to physical laws which are orderly and rational, man often behaves irrationally in ways that do not serve his own best interests. Senator Joseph McCarthy, Hitler, and Nasser are dramatic examples.

I am not the first to recognize the irrational component in man. Said Laplace many years ago, "Our passions, our prejudices, and dominating opinions, by exaggerating the probabilities which are favorable to them and by attenuating the contrary probabilities, are the abundant source of dangerous illusions."

Logical necessity does not always lead to logical behavior. I take this as my third constraint since there seems to be no reason to assume that man will change markedly in this respect anytime soon. Indeed, if we look at some recent experiments with rats we are forced to conclude that irrational behavior is a genetically-transmitted side effect of some other trait biologically evolved to adapt fauna to their environment. In the rat experiments to which I refer, the animal can electrically excite a compulsive (pleasure) center of his brain by depressing a lever. This he does to the total neglect of food and sex, at rates up to 5000 times per hour, until he falls over exhausted.

Some of you may know of the very fine work by Dr. Bernard Saltzberg on brain functioning and behavior. He tells us of a woman dying of cancer whose extreme pain was not alleviated by any drug. Yet with just one short electrical stimulation of her brain each day, she forgot her pain, her eyes brightened and she reported feeling well until the day of her death.

The jungle instincts which still govern man after a million years of cultural but little biological evolution do not include a high tolerance for ambiguity. He would not have survived his enemies had he been willing to await proof of their hostile intentions. Rather he learned to fight or to take flight, but he seldom suspended judgment.

Today, in the face of intellectual uncertainty, even when it is not accompanied by any overt threat to his safety, man accepts magic as a way of life. I see no imminent decrease in such beliefs, and they are certainly not correlated with I.Q.

Last year and the year before I taught a course called "The Boundaries of Modern Science," in which we devoted about two weeks to antiscience and magic. All the students had better than a B average. Yet, I found a large proportion who believed in flying saucers, extrasensory perception, water witching, fortune-telling, and other forms of magic.

Berelson and Steiner in their unique summary, *Human Behavior*,[5] give us an insight into what reinforces a belief in magic. I cite a few of their theorems about learning, referring you to their book for the supporting evidence:

1. A 100 per cent reinforcement schedule is usually the quickest way to establish new behavior or to increase the frequency of a response; that is, learning proceeds most rapidly when every correct response is reinforced. Responses learned on such a schedule also extinguish most rapidly.

2. Learning persists longer with a schedule of partial reinforcement than with 100 per cent reinforcement.

3. Punishment on a 100 per cent schedule depresses the response rate more quickly than intermittent or partial punishment, but the effects of partial punishment last longer.

Next I cite the theorem that is so illuminating to our present discussion of irrationality:

4. Response rates become more stable, and behavior more resistant to extinction, if reinforcement is inconsistant; that is, if it varies in rate or in interval.

Purely by chance, a magical rite is occasionally confirmed. Sometimes the witch finds water, some flying saucer reports are not identified with known natural phenomenon. These all strengthen belief.

An adjunct theorem further explains the persistence of man's irrational belief in magic:

5. The less frequently reinforcement occurs, the greater the tolerance for failure . . . superstitions persist largely because they work infrequently and sporadically, not in spite of that fact.

It should not be concluded that all irrational behavior is undesirable. For example, in "The Sociology of Sociability," George Simmel writes:

Since sociability has no ulterior end, no content, and no result outside itself, it is oriented completely about personalities. . . . Rationalism, finding no content there, seeks to do away with sociability as empty idleness. . . .

Simmel takes his stand with the angels by concluding:

everyone should have as much satisfaction of this impulse as is consonant with the satisfaction of the impulse for all others.[6]

In view of recent experimental findings that babies need fondling to survive, and of other similar data on lower animals, it is probably irrational to classify sociability as an illustration of irrationality.

Says William C. H. Prentice, former Swarthmore Dean and now President of Wheaton College:

> The future of our world, if it has a future, lies with people who are basically unselfish, people who think of themselves infrequently and who, when confronted with a clear choice between self and others, often make the choice in favor of the others. . . . People who are idealistic, setting standards and goals for themselves and for the world which may transcend realistic achievement but which can nevertheless fix the direction of a life. People who thrive on effort and hard work and sacrifice.[7]

"In a perfectly rational society," says Don E. Kash of Purdue University, "policy would be made something like this: Goals would be articulated. Various strategies for achieving those goals would then be defined. A cost-benefit analysis of the various strategies would follow. And the final step would require taking action along the lines of the selected strategy."[8]

We know that this process guides only a relatively few personal lives, corporations, and countries. The drive for individuals and groups to get and keep power is overriding.

The Ambiguity Constraint

Man is bound to blunder, to describe as timeless a principle whose life is limited, to claim universality for a precept whose geographical, ethnic, or language boundaries he sees only dimly or not at all. This leads, then, to a very valuable operational guide: Science as a way of discovering error, not truth.

Science tells us to behave as if all of our philosophic conclusions are relative, tentative, approximate, and in need of repair, as indeed they are.

This ambiguity constraint causes psychological pain. Many people, possibly most people, want answers so badly that they'll accept one even known to be wrong, in preference to no answer. How many times have we all heard someone say, "Let's drop the bomb and get it over, even if I do die myself in the retaliatory strike."

We don't know the answers to many questions and may not at any time in our life. We must, therefore, develop a tolerance for ambiguity and teach our children to avoid the anxiety that lack of such tolerance engenders. We may take comfort in the many answers that the human race has found to past ambiguities.

Law as a Constraint of Cybernetic Cultures

There is space to discuss just one more constraint and this I choose to be law. The unique contribution of Hammurabi (1955 B.C. to 1913 B.C.), who wrote his code of laws nearly 4000 years ago in order to promote social justice, was to raise normative human relations to a high level of abstraction, independent of particular individuals.

Now law is the written, long-time average result of social force applied in conflict. It is the distillation of experience and knowledge into wisdom. A necessary constraint is, therefore, the requirement that there exist a set of time-stable and generally approved rules governing interrelationships among people. Furthermore, it is essential that each member of the group understand the rules and obey them, on the whole. Yet these rules can never be all encompassing. Society must continue to live on unverified assumptions. Otherwise we couldn't eat hamburger or stew.

This constraint appears to be trivial until we remember that the theoretical goal of many utopias, and of Stalin's Russia, was to eliminate all law.

I leave it to you to decide whether this constraint is relevant in a discussion of Chicago police action during the Democratic convention of 1968.

Our mutual dependence upon each other for the necessities and pleasures of life is not the only reason why our caprices and even some of our freedoms must become increasingly abridged. Within the past few years every spot on earth has become accessible to every other by a guided missile. Society just cannot afford to live so uncertainly. Without touching upon the mechanisms by which controls over our behavior will be established, it is a logical consequence that the price we must pay for global life insurance is restriction on the display of our sadisms and masochisms.

Since mistakes inevitably occur, the rules of social organization should provide for peaceful recall of decisions. It must be possible to amend laws within a time period that is neither so long that tensions grow to disrupt the society nor so short that people fail to learn and obey the laws. The U.S. Constitution seems to strike a happy balance in this respect. One of the reasons that Robert Owen's New Harmony failed in 1828 was its lack of a stable set of laws. After the community hammered out a constitution over a period of almost a year, during which Owen was in England, he arrived at the colony and promulgated a different constitution.

It must always be illegal and, hopefully, impossible to abrogate this amending process down a one-way street. A legislative body which dissolves its constitution, as the German Reichstag did in 1933, ensures revolution as the only remedy to the absolute power which corrupts absolutely.

VI. Summary and Conclusions

Summary

In this article I have reviewed some of the characteristics of utopias. Their rigid control and values, isolation, small size, external rules of behavior, and lack of experimentation doom all such attempts to failure.

The opposites of these characteristics in cybernetic cultures encourage one to hope that they can exist in large size and for an extended period. Indeed, the United States of 1969 meets most of these requirements and we may claim, then, that we are living in the most successful culture yet developed by man.

I assume that man will continue to build utopias and learn to convert these to cybernetic cultures in spite of the cynicism of people like New York's former mayor, Jimmy Walker, who defined a reformer as a "man who rides through a sewer in a glass-bottomed boat."

If we are to build cybernetic cultures we should make full use of the extensive history of utopias and of the history of military, religious, and business management. Out of such history I draw certain principles, or call them constraints in the language of simultaneous equations. As we seek to improve what is already so good, by the standards of our own history and of other countries today, we should conserve these constraints. From an undetermined larger number, I have presented several for you to ponder, as follows.

A viable cybernetic culture must:

1. Encourage a moderate level of diversity.
2. Provide universal education as an antidote to the original ignorance of all persons.
3. Expect and make provision for irrational opinions and actions.
4. Increase its citizens' tolerance for ambiguity.
5. Maintain a living body of democratically determined law.

Conclusions

With the specter of massive and permanent starvation facing most underdeveloped countries starting about 1975, and in view of our nuclear confrontation with the communists, it is to America's self-interest to help build cybernetic cultures elsewhere in the world. Since the blueprints for these are not known we need all the research wisdom and management talent the country can assemble to stay ahead of catastrophe. It is not at all certain that we shall succeed.

Someone once said that a man who deludes himself that America's problems have solutions is called a philosopher, and when he tries to convince his neighbors of the same delusion he's called a candidate.

Being a "cockeyed optimist" from the North Pacific country, I think that such delusions are magnificent.

Is there any better ideal for us to suggest to the next generation?

REFERENCES

1. Marie Louise Berneri, *Journey Through Utopia* (London: Routeledge and Kegan Paul Ltd., 1950), p. 40.

2. G. Evelyn Hutchinson, "Turbulence as Random Stimulation of Sense Organs." *Cybernetics,* H. Von Foerster, ed. (New York: The Josiah Macy, Jr. Foundation, 1953), p. 155.

3. Ronald L. Lee, "A Threat to Individual System Functioning." *The Rocky Mountain Social Science Journal,* Vol. 4, No. 2, October 1967, p. 4.

4. Arthur M. Schlesinger, *The Bitter Heritage: Vietnam and American Democracy 1941-1966* (Boston, Houghton-Mifflin, 1966).

5. Bernard Berelson and Gary A. Steiner, *Human Behavior* (New York, Harcourt, Brace & World, Inc., 1964) pp. 151-157.

6. George Simmel, "The Sociology of Sociability." *Theories of Society,* Parsons *et al.,* eds. (New York, The Free Press, 1955) pp. 157-163.

7. William C. H. Prentice, Address on the Occasion of Father's Weekend at Wheatley College. Norton, Massachusetts, 1965.

8. Don E. Kash, "Research and Development at the University." *Science,* Vol. 160, June 21, 1968, p. 1313. Copyright by the American Association for the Advancement of Science.

9. Samuel Clemens, *Mark Twain's Notebook* (New York, Harper & Row, 1935), p. 381.

Brain Research—
How to Look at Unobservables

DONALD O. WALTER

Brain Research Institute
University of California
Los Angeles, California

My talk is likely to be disorganized today—perhaps as an image of our society. It is difficult to pull things together, not only in describing brain research, but also in trying to express the analogies that we are said to be searching for; so I'll give you a collection of *hors d'oeuvres* from which you will have to make your own nourishment.

Professor Hughes' celebration of the positive aspects of cybernetics as applied to society leads directly into some quotations with which I want to start, which don't have anything to do with brain research directly, but are part of my motivation.

The first one comes from a book review by Alvin Gouldner of Washington University, which recently appeared in *Science;* the book is on sociology and is edited by Talcott Parsons, who spoke to us last year. Gouldner dislikes it in general, and he gives supporting examples: "Charles 'Tilly's essay on urbanization presents another important case in point. 'No country,' complains Tilly, 'has a social accounting system allowing the quick, reliable detection of changes in organizational membership, kinship organization, religious adherence or even occupational mobility.' From Tilly's standpoint as a research-oriented sociologist, this is a bad thing. Yet, what kind of country would it be that would have such a relentless, quick, reliable, all-embracing system of information about its population? Surely, it would be a country in which the potentialities (at least) of the most complete totalitarianism were at hand. Doubtless Tilly would reject such a society as quickly as I. What he and others fail to see, however, is that many of the conventional methodologies of social research premise and foster a deep-going authoritarianism." [1] Later Gouldner says, "Such social sciences will be mindlessly ready to buy increments of information at the cost of human dignity and freedom. In short, then, from its substantive assumption that human beings are 'in fact' the raw materials of independent social structures, to its methodological assumption that men may be treated and studied like other 'things,' there is a strong repressive current in sociology, as in other social sciences, a current that congenially resonates the impulse of any modern political elite to view social problems in terms of technological paradigms, as a kind of engineering task." Such a view, I think, is very congenial to cyberneticists. We are gathered here to view society as a sort of engineering task, and I think we have to continually be asked by others, if we do not constantly ask ourselves, what sort of a politics and what sort of a society would emerge from the information systems or the control systems or the cybernetic whatnots that we cheerfully discuss.

Another quotation that I want to read you, to show that our cybernetic society does have tendencies that way, I take from the ASC Newsletter, written by one of our distinguished vice presidents, Dr. Hammer; it says in part, "Authorities on the subject assure us that by the end of this century, electronic systems will affect practically every aspect of human endeavor. Every individual will have then at his or her disposal a vast complex of computer services." [2]

Now, *every individual* is a lot of people. I wonder whether he means every individual like us gathered here today, the poor, Negroes (of which there is only one here), women (of which there are a few) or people under thirty (of which there is practically no one here). I don't think he really means *every* individual. How about un-Americans? How about Latin Americans? Gouldner's review has another related quote: "Reinhard Bendix also reassures us that in modern society the words 'ruler' and 'ruled' no longer have 'clear meaning.' Presumably this is so because people now exercise 'control through periodic elections,' and 'the fact that every adult has the vote is a token of the regard in which he is held as an individual and a citizen.' The franchise, Bendix tells us, has been 'extended.' One wonders if that is how the matter would be put by those who were arrested, beaten, and killed in the struggle to register Blacks in the South—would *they* see what had happened as an 'extension' of the franchise?" In other words, as cyberneticists, it is so much easier and pleasanter for us to deal with systems which are stable, or in control, or well-behaved, and it's so easy to take an engineering approach which leaves out the really unimportant or irrational or whatnot side of things that is hard to formulate. We offer you an approximation, we cyberneticists often seem to be saying. What do you want? Engineering and science are always approximations. How can it be bad to make an approximation which leaves out things that are usually unimportant? This two-headedness, of cyberneticians doing cybernetics vs. doing ordinary things, is the burden of some of my distress with the proposed cybernetics of social systems.

Determinism and Chaos

Now, to try to suggest a positive possibility toward solution of some of these problems, I turn to one of the fundamental difficulties in the applications of science to life—by which I mean both life science, which I'm a practitioner of, and sociology and the like: the difficulty of bringing together the idea of determinism or predictability (which naturally goes along with a nice, closed-form, mathematical model which works pretty well most of the time), and the feeling which is often expressed in the phrase free will, or related expressions. How is it that there can be important indeterminism in people or in societies? In isolated models, in models of *parts* of systems, we can say that the randomness or the stochastic process or the jiggling comes from outside; but when we start to have a world society or when we start to make a model of a *whole* thing of any kind, then there are no external sources of randomness (or of order). But we have this model, coming from old-fashioned physics or old-fashioned engineering, of determinism; if we don't have external sources of randomness, then the system somehow must encompass randomness inside itself. The trouble is that randomness seems to be meaningless, to preclude meaningfulness. If you are trying to make a model of a whole system which is so whole that it doesn't have external inputs to which you can refer either order or randomness, how do you avoid thinking either in terms of meaningless determinism or else meaningless chaos? What is there in between?

This problem came to me most importantly in terms of how to think about the brain as a piece of physics and, yet, as a part of ourselves which is not completely determined but is not completely chaotic. Where is the stable foothold midway between determinism and chaos? The particular image that I had, which I think gives us the way of thinking about this foothold and building on it, is formulated in terms of very explicit models of brain function, which in a way are irrelevant to the general idea of establishing this philosophical foothold. I think, however, that I can best explain the image to you in that setting; but I ask you, please, not to attack the particular mechanizations that I'm proposing, because they are not as important as the fundamental image.

Suppose that the brain waves which I spend most of my time studying go along with behavioral situations, with changing life experiences, and that when the situations are repeated (as in a conditioning experiment, where we impose a lot of regularity on a person or an animal so that we can get something to study that doesn't vary as much as freely moving behavior does) the waves, too, will be repeated to some extent, though not completely. Now consider the effect of these waves on the microstructure of the brain—for instance, on the chemical flavors of RNA that are produced in various parts of various nerve cells. (Any other process which leaves a physical trace and which could be influenced by these waves is adequate as the substrate of my theory.) And, these supposed flavors of RNA may have effects on the nerve cell's membrane, or on the synaptic interaction between cells, or the like. Here arises a very important problem: how does this repeated activity go about getting coded from the large (the waves and the transmission of impulses) into the very small? Not only recorded and coded in such a way that it can remain physically stable, but also so that it can be recalled, so that there is some way of reading the punched molecule once you get it laid down. This can't be prewired, this can't be in the blueprint, there isn't enough space, I believe, in the genome to specify all of this stuff; how is this coding accomplished in such a way that it can be and is often decoded?

This intermediate coding problem is one which is easier for neurophysiologists to leave aside than to address; but I am going to offer you a very speculative suggestion: that if there are interactions of the kind I was suggesting, between the gross brain waves and the microchemical factories, they can be *random* in one sense. Now, random is a very much maligned and misused word. The sense that I mean is, that it is not necessary to specify beforehand whether a given wave shape and timing and pattern will stimulate or depress a given tiny area in one little cell out of the 10^{11} nerve cells plus 10^{12} other cells inside the brain. This we can think of as random, in the sense of being undetermined, not laid down specifically in advance by some other agency. And the effect of this differential stimulation or suppression of production of different flavors of RNA on the membrane (or whatever is going to allow this memory to be recalled) can also be random in that same sense of not having to be laid down in advance. There should be some effect, but it doesn't have to be laid down in advance whether it's to be more of the same or less of the same. So, it looks like we are stuck on the chaotic horn of the dilemma, but I think not for two reasons. First, because at the microlevel these processes are *unobservable* by external electrodes or wires or whatever we might plant inside the brain, because we'll disturb them too much;[3] *you'd* be disturbed by a wire implanted in *your* brain. We disturb them too much, they are unobservable—I think this is an important fact, but I won't pause to argue it at the moment. But, as a second essential factor, there is a *natural selection* operating on the waves as they tend to repeat themselves—a cumulative form of natural selection, because traces of the previous waves have been left behind; thus the outcome—which waves will be emphasized, amplified, passed on, and continued and which will be suppressed or discouraged—is subject to natural selection. I mean the phrase natural selection to be understood in the broad sense of the theory of evolution by natural selection, and I mean to suggest the importation into the brain, at this microlevel, of all of the mechanisms of competitive exclusion by competition for resources and the like that we are told about by population genetics. Furthermore, this natural selection continues at higher levels of organization, where it merges into conscious or intentional selections by me, the inhabitant of this brain, or you, the inhabitant of your brain. And yet, there are thoughts, I believe, and conscious choices (or semiconscious choices) which are also not really observable because, if I stop to tell you what I'm thinking, it interferes with my thinking that same sort of thing; so we still have an indeterminancy principle of observation. But they are all subject to selection—natural selection at the very smallest scale, which merges gradually into a conscious or intentional selection of what I will do next. And some of this is observable to us, some of it even observable to outsiders. Some of it is predictable, some of it is not. But notice, as an additional bonus, that I have implicated, along with the idea of natural selection, the still marvelous fact that you and I exist as the (no doubt) peak product of evolution to date. It has been elucidated to some extent how it is possible for a process like natural selection to concentrate improbability, to concentrate selective information;[4] and so I'm suggesting that even though we don't have a full understanding of natural selection in evolution, especially on the macro-scale and concerning the emergence of new species, that these problems can profitably be put into an analogy with each other and thereby, I think we have at least a philosophical ground for a position midway between chaos and complete determinism in the brain.

A Social Analogy

When I try to explain the brain to really lay audiences (not like present company) I am often at a loss for an image, an image which conveys what I feel about the complexity in the hierarchial organization and the unobservability and the complication of the brain, but, some years ago, I had decided that really a fairly communicative analogy would be to say that the brain is like the American economy and the American social structure. There are variables in both systems which are very easy to measure and other variables which are very hard to measure or observe. In the brain, it's terribly easy to measure voltage waves that go up and down, but it's terribly hard to measure something like impedance, locally. And analogously, in the social structure and the economics of the United States there are quantities easily measured and others very hard to measure. I use the United States not only for ethnocentric familiarity, but because I want to counteract the argument that some people make, that we should understand the simplest organism before we try to understand the brain of man; I think this is dandy if you do indeed understand the simplest organisms and know how to make the extrapolation, but both of these are necessary. This is like saying we have to understand the economics and social structure of the Canary Islands and then of Ceylon before we can come to understand and do anything about those of the United States. This social metaphor also seemed appropriate to me as something of an antidote for the computer model of the brain—brains are a little tiny bit like computers, but not very much so. And it's an exaggeration of simplicity

32 — DONALD O. WALTER

to analogize a brain cell and its function to a flip-flop or a one-shot and its function. It's admittedly an exaggeration of complexity to treat a brain cell like a human being, to say that the brain is like the United States, a society composed of cells like people; but I don't think it's any worse exaggeration than the computer image, and it brings in the supplements that need to be brought in. Now that I've convinced you (no doubt) that there is a foothold between determinism and chaos, then perhaps there is something of the same kind in modeling society.

To come back to my opening disordered quotations: I want to urge cybernetic modelers to include in their models something in the nature of freedom, you see. It's easy to say, "Yes, we must include human dignity and freedom in our utopias," and it's important to say so, but it's difficult to know how to proceed to do that, because when we put down a model it is generally a relatively simple one. At any rate, it's closed, it's somewhat well-behaved, it has only one or two orders of size in it, and it's likely to leave out disturbing things like unpredictability or disorder or freedom. I don't know how to get dignity into a computer model; but I urge you to accept that here is a feasible mechanism for at least incorporating the possibility of something between determinism and chaos, and if that *can* be incorporated in a model of the cybernetics of society, I'll be very happy.

REFERENCES

1. Alvin Gouldner, "Disorder and Social Theory." *Science, 162,* October 1968, pp. 247-249. Copyright 1968 by the American Association for the Advancement of Science.
2. Carl Hammer, *ASC Newsletter,* Vol. 1, No. 4, October 1968, p. 3.
3. D. O. Walter, "The Indeterminacies of the Brain." *Perspect. Biol. Med., 11,* Winter 1968, pp. 203-207.
4. M. Kimura, "Natural Selection as the Process of Accumulating Generic Information in Adaptive Evolution." *Genet. Res., Camb., 2,* 1961, pp. 127-140.

What Is Normal to Physicians And Society

STEPHEN L. SHERWOOD

Stanford University School of Medicine
Palo Alto, California

The growing number of computers used in biology and the growing areas of their use compel us to rehearse many of the definitions, parameters and scales which until now have resided in our heads as subsumptions. Many of these are there in the form of unquestioned tradition. Now that we are handing over to computers many of our chores and, indeed, many of our responsibilities, we must be certain that the nature of our questions is within the scope of the computer's program and within the competence of the program's design.*

The science of life, growth, production and reproduction have converted farming from a craft into a highly technological industry which has, as its efficiency standard, the market. Medicine, now, is also being converted into an industry, but the standards of success are less well-defined. Hospitals are becoming "people-repair shops."

Editor's note: The idea of handing over responsibility to computers is discussed at length in *God and Golem, Inc.* by Norbert Wiener (MIT Press, 1964).

What *Is* Normal?

Common usage has it that normal means "conforming to standard, regular, usual, typical; . . . average or mean of observed quantities; usual state, level, etc." [1] This description in terms of statistical probabilities (in biology at least) goes back to Bernoulli, Laplace, Fechner and Gauss.[2] It is clear that qualities such as excellence, beauty and tallness are as improbable as inferiority, ugliness and stuntedness, and so are heroism and cowardice. Galton [3,4] not only examined the inheritance of mental traits in terms of probabilities, but also attempted to correlate these with physical traits as inherited. To him it was a matter of eugenics. He came to the conclusion that "the *goal* of human effort is not heaven, but superman." [5] Superman is not normal in Gauss' sense.

The word "goal" is, of course, the cue: goal implies teleology, such that the trend, or aim, or "goal-seeking" behavior has a teleological vector away from the mean and the norm.

As far as I can tell, at any one time there have been, and are, at least four recognized determinants of what is normal, and hence acceptable, in the world of medicine: the patient's desire; the opinion of society; the physician's rule; and the arithmetic of distributions.

It is fitting here briefly to review the history of medical efforts.

Galen, in the second century A.D., was a great man and logician. Probably one of the first experimental physiologists and dissectors, he discovered facts about the nature of living beings either then unknown or only guessed at. It was the misfortune of Galen's findings that a contemporary of his, Celsus, who had Platonic notions even though he was moderately anti-Christian, became so popular as to make a dogma for medicine out of Galen's teachings. The consequence of this was that original investigation was paralyzed until the appearance of a man, Theophrastus Bombastus Paracelsus (von Hohenheim), a Swiss, who ran afoul of his own society. Essentially an alchemist, he started lectures to students in German, the vulgate, and before his audience burnt the books of Celsus.

He also burnt the books of Avicenna, the Arab, the latter's writings less deserving of such a fate than those of Celsus. Much of his work was mystical; he professed that man's life was an extract of all beings previously created and that all ills derived from a separation of elements that should have been compounds. But also, man must have a soul created by God. In passing, it may be noted that Paracelsus was given a living by the Archbishop of Salzburg, who was by no means averse to good living.

At about the same time or a little later, the great Italian anatomists, from Leonardo da Vinci to Morgagni, revived medicine as a branch of original and real investigation with their illicit dissections and inquiries. Let us observe that in the same era William Gilbert of Colchester wrote his book, *Of the Magnet.*[6] As court physician to Queen Elizabeth he was, to my knowledge, the first real model-maker. He alluded to the behavior of heavenly bodies by way of lodestones. Of course, in a then Protestant England, he did not have the impediments of Galileo, who lived under the Pope's eyes.

The last few paragraphs may seem only tenuously connected with the question of "what is normal." But all these men were in the business of discovering laws for functions they otherwise could not understand. In the Renaissance and the following age of enlightenment, men like Erasmus of Rotterdam (or Gouda) and his group of philosophical explorers fought the "Finstermänner," the obscurantists. As apologeticists in theology, and by calling the shots on dogmatic tradition, they enabled those who would to find the truth by whatever means were necessary and adequate. They taught by argument rather than precept. Only after their work, and that of Descartes and Hume, could experimenters flourish. Let us remember that only after extensive philosophical search could Descartes investigate the physics of his Opticks.

This, then, was the triangular relation between the mores of society, the desires of the patients and the physician's rule. From that time on the great physicians advanced through to our own times by trial and error and search for facts. But let it not be forgotten that after Boerhaave, court physician to the

Empress Maria Theresa, had opined, with success, that a tickling of the imperial vulva might overcome her infertility, Sir William Osler, without fear of ridicule or lack of logic, could state in his great text of 1909 that only clean living was a remedy for syphilis.[7] We may observe that he was a born Canadian, made his way through McGill, Pennsylvania and Johns Hopkins to Oxford, all that being top-hat, frock coat and striped trousers.

Some people, especially Mediterraneans, are born "successful"—the French word *re-ussie* is a little better and more precise. The classicists clarify, they resolve; others, the romantics, live with, and in, a perpetual storm and urge—the old Nordic *Sturm und Drang*—which is never resolved, unto self-destruction. Compare the talents of Botticelli and Rembrandt, Rossini and Wagner, Mark Twain and Dostoyevsky, or Gastaut and Eccles, or the creeds of St. Francis and Martin Luther. The boundaries are cloudy; Nordic mists do sweep south and pellucid airs waft north. Compare Leonardo da Vinci and Vermeer Van Delft.

Most of us have taken refuge under the bell-shaped curve to define the norm;[8-14] others have or are looking for multiple variants. The need to come to grips for the meaning of "normal" [15-18] was brought home to me by the following. First, the need to define "normal" for a computer; it has a program which decides, for instance, that certain brain waves which are normal for a child or a man asleep are not normal for a grown man awake; second, I overheard one physician's remark to another: "For this disease, such a high fever is normal"; and third—and this is the key—an astronaut's report: "All systems normal."

Now, a system is a collection of interacting units, put together for a purpose. The word "system" is defined by the *Concise Oxford English Dictionary* as a "complex whole, set of connected things or parts, organized body of material or immaterial things . . . ." In biology the word "system" can well stand for an organ; such an organ fulfills a purpose. A collection of organs whose function, together, fulfills a purpose may be termed an organism. This implies that in the definition of normalcy the purpose of the system, organs and organisms must be understood. Teleology must be part of the analysis.

I am not quite certain what a large system is. To an electron an atom is a large system; to an atom a molecule is a large system; to a molecule a living cell is a large system; and to an organism a society is a large system.

Now the physician is concerned with all of those. Between him and his patient a decision has to be made on what is normal. How, then, should the machine be instructed to assess and assay biological data?

Gauss' curve tells us that members of a given population have properties in common, that they will conform to those properties in various degrees and that the largest number of individuals in the population will be found amongst those who conform most closely to the characteristics defining the criteria, i.e., that which is conventionally the standard of normal.

This paper is the expression of an attempt to demonstrate the distinction between average and normal for physicians and society. The thesis is that "average" is a well-defined arithmetical description whereas "normal" is not. "Normal" contains teleological constraints; embedded in the notion of normal is a purpose and hence an intent, a goal seeking behavior. In theological terms, it would imply free will which is not normal because it does nothing that is determined by the average, or Gaussian, distribution. Neither heroism nor cowardice can be called normal because the hero intends to risk his life (teleologically) for the good of his faith, and the coward intends to save his life, no matter what the cost to others. We may mention at random, Julius Caesar, Dostoyevsky and one outstanding contemporary orchestral conductor, who were abnormal in two ways: First, they excelled in their goals, and second, they were in the class of what contemporary medicine calls "chronic brain syndrome," with the usual rider "un-differentiated." It would be interesting to speculate what this description would have done with Henry VIII, Nietzsche or Schumann, or the man who led Britain and her allies to victory in 1945, and who is said to have liked a fair amount of ethanol in the paradigm of brandy.

We may also personally recall the case of a Welsh Guards officer, one half of whose brain had been carried away by gunshot. It took three or four surgical operations to patch up his skull; following this he managed to graduate from Oxford and to obtain an appointment in the Bank of England. Is this normal? (The appropriate discussion may be found in Keynes' *A Treatise on Probability*.[19])

There is a further question to be considered: Masters and Johnson, as quoted by Saunders,[20] state that they abandoned work with prostitutes because they were abnormal. Now, what was the norm: behavior of prostitutes or of the general populace? Admittedly, the greater part of the study concerned itself with people of some higher education and financial solidity. What is normal for them may be very abnormal for prostitutes.

We have, then, the following possible descriptions of normal.

1. Conformity with Gaussian distribution:
   a) the curve may be peaked
   b) it can be blunt
   c) the more it is peaked, the greater the significance of data falling under it.

2. The significance of measurements:
   a) how valid are the data respecting significance?
   b) criteria can be applied which will optimize meaning
   c) meaning, for the purposes of this paper, is that information which enables one best to predict results from a concatenation of circumstances
   d) obviously, this implies an understanding of the circumstances.*

---

*Editor's note: A good book for the nontechnical reader who wants to gain information and understanding of information theory is "Symbols, Signals and Noise" by J. R. Pierce, (Harper & Bros., 1961).

A consequence of this is the application of Shannon's theorem $H = \Sigma p_i \ln p_i$.*

This theorem applies to closed systems only, which is comparable to Boltzmann's theorem (quoted by Schrödinger: *Heredity and the Quantum Theory*). "Significance," then, is a direct derivative of meaning and information in Shannon's sense: for example, one may record for twenty-four hours the brain waves of a patient: if a telltale three seconds worth of "spikes and waves" occurs, it is sufficient information to classify him as an epileptic and therefore abnormal.

We have now two different statistical results. First, that which shows that most members of a class do *not* all have a given characteristic, and, second, that those who do have it depend not only on the improbability but also on the particular nature of the deviation from norm. This double standard can readily lead to contradictions. An efficient politician may be said to be an "operator" or an able statesman; a brave man may be called foolhardy or a hero; and the donor of money may be classed as a wastrel or a benefactor. Vice and virtue may thus be both equally abnormal, and the same characteristic can be judged differently, depending on the set of criteria. Hence, "normal" must be referred to a set of values beyond the purely statistical measure of occurrence. Myxomatosis, some years ago, wiped out all but a few rabbits in the United Kingdom: up to that time they had become a plague to farmers (abnormal); rabbits contracted a fatal disease (abnormal); the rabbit population was decimated (normal for farmers); foxes and birds of prey had to turn to snails, larger insects, etc. (abnormal for predators) until some resistant rabbit strains increased their numbers (normal for rabbits and predators, abnormal for farmers).

Hence, the measure of information may differ not only in value, but also in its nature, which in turn depends on such overlap as may exist between different closed systems such as the information exchange between rabbits, a virus, predators and people.

---

* **Editor's note:** This formula is a definition of negative entropy which Shannon used as a quantitative measure of "information."

It is not normal to attempt to climb Mt. Everest, or to wish to build the Taj Mahal; but, once one has made such a decision, the execution of such an endeavor becomes normal.

The first pointer in the direction of teleology being part of the definition of normal came from the anthropologist, Sir Francis Galton.[3-5] He applied statistics and methods of probability to the family backgrounds of eminent men, and he can be said to be the founder, after the Rev. Malthus, of eugenics.

In the same school, Penrose,[22] after Galton, statistically analyzed the probability of mental deficiency and other handicaps for children born of certain families. He was impressed enough by his results to study inheritance patterns and their results in a relatively simple model, namely the coloring of *Drosophila melanogaster,* or the common fruit fly, arithmetically, as indeed had Mendel studied plants a long time before him.

Since then mathematics, and in particular statistics, have become a prime tool in the study of what is normal. Pareto [23] with his studies of the distributions of income, J. B. S. Haldane [10] with his biometrics and Menger [24] have pointed to possible ways so that we may use exact methods in arriving at a definition of what is normal. As a society we must spell out the purpose of normalcy and the transitional probabilities of these our conclusions.

Should we aim for longevity? George Bernard Shaw lived much of his life on nuts and grass. He suffered much from sexual impotence and recurring infections. His artistic accomplishment and his longevity are not in doubt. Should we breed for athletic prowess as a norm? Athletes in general are not usually outstanding in their latter years for attaining what society calls high levels of achievement or for longevity. Since the first step in 1948 toward national and universal health services in the western world, health is indeed and inexorably becoming an industry. Obviously, as an industry for mass consumption, it cannot be at a Rolls Royce level. I suspect that the norm will be established by peoples, governments and their wealth, and climate. The information exchange, no doubt, will obey a law such as the entropic $H = \Sigma p_i \ln p_i$. The adjustments toward the norm will be done by appropriate error-correction procedures.

We can now close the circle around "what is normal." Normal is not only that which occurs most often, but that which is referred to a given set of defined aims.

The results of these considerations impose the following constraints on the search for knowledge, or "research."

1. The concept of "normal" is not absolute as a binary alternation of zero or one.

2. It must be referred to a frame of reference, which must needs contain a purpose, be teleological.

3. Hence, it must be in part related to time, place and circumstances.

4. Finally, in view of the above three points, it has to do with homeostasis, the trend toward stability. In more contemporary terms, the normal is the result of servo mechanics— of course within bounds: if the error exceeds the capabilities of the governor, abnormality is the result. Our world has perhaps reached such a critical state in recent years.

REFERENCES

1. H. W. Fowler and H. G. Le Mesurier, eds., *The Concise Oxford English Dictionary*, 3d ed. (Oxford, Oxford University Press, 1934).
2. J. R. Newman, ed., *The World of Mathematics*, 2 (New York, Simon & Schuster, 1956), p. 1157.
3. Sir Francis Galton, *Inquiries into Human Faculty and Its Development* (New York, Macmillan, 1883).
4. Sir Francis Galton, *Hereditary genius: an Inquiry into Its Laws and Consequences* (London, Macmillan, 1869).
5. E. G. Boring, *A History of Experimental Psychology*, 2d ed. (New York, Appleton-Century-Crofts, Educational Division, Meredith Corporation 1950), p. 483.
6. William Gilbert of Colchester, *De Magnete*, trans. by P. Fleury-Mottelay (New York, Dover, 1958).
7. Sir William Osler, *Principles and Practice of Medicine*, 7th ed. (New York, Appleton, 1909), pp. 278-279.
8. R. W. McCammon, "The Concept of Normality." *N.Y. Acad. Sci., 134*, art. 2, pp. 559-572.

9. Ward C. Halstead, *Brain and Intelligence* (Chicago, University of Chicago Press, 1947), pp. 6ff.
10. J. B. S. Haldane, "Time in Biology." *Science Progress, 175,* 1956, pp. 385-402.
11. A. R. Feinstein, *Clinical Judgment* (Baltimore, Williams and Wilkins, 1967): diagnostic taxonomy, pp. 72, 89, 103; objectives of treatment, p. 231; indices of criteria of therapeutic response, p. 247; medical records, p. 264; objectivity and precision, p. 306; mathemedical construction, p. 173; normality, criteria for, p. 330; criteria (distinguish arthritis, arthralgia, myalgia), p. 331; specificity of results (pathognomonic), pp. 333-334.
12. G. L. Engel, "A Unified Concept of Health and Disease." *I. R. E. Transactions on Medical Electronics. ME,* 1960, pp. 48-57.
13. Carl Black, "A Computer Approach to the Parametric and Nonparametric Description of Distributions and Their Subsequent Normalization Using a Polynomial to Obtain Normalized T-Scores." *N.Y. Acad. Sci., 134,* art 2, 1966, pp. 538-540.
14. G. L. Engel, "What is calculus of variations and what are its applications?" in *The World of Mathematics,* T. R. Newman, ed. Vol. 2 (New York, Simon & Schuster, 1956), p. 886.
15. Ernest Simonson, "The Concept and Definition of Normality," *N.Y. Acad. Sci., 134,* art. 2, 1966, pp. 540-558.
16. Warren S. McCulloch, *Cybernetic problems of learning in Conditional Reflex, 2* (1), 68-76. J. B. Lippincot, 1967.
17. J. B. S. Haldane, "Les Aspects Physico-Cliniques des Instincts" in *L'Instinct dans le Comportement des Animaux et de L'Homme.* (Paris, Masson et Cie), pp. 545-559.
18. A. Sollberger, *Biological Rhythm Research* (Amsterdam, Elsevier, 1965).
19. J. M. Keynes, *A Treatise on Probability* (London: Macmillan, 1921).
20. M. K. Saunders, "The Sex Crusaders from Missouri (Masters and Johnson)." *Harpers Magazine 286* (1416), 1968, p. 48.
21. T. R. Newman, ed., *World of Mathematics, vol. 2* (New York: Simon & Schuster, 1956), p. 992.
22. L. S. Penrose and C. K. Myers, "Method of Preliminary Psychiatric Screening of Large Groups." *Amer J. Psychiat. 98,* 1941, pp. 238-242.
23. Alfredo Pareto, quoted in *The World of Mathematics,* J. R. Newman, ed., 2. (New York: Simon & Schuster, 1956), p. 1200.
24. K. Menger, "Variables, Constants, Fluents." *Symposium on the Logic of Variables and Constants, Proc. Nat. Acad. Sci., U.S.A., 39,* 1953, 956-961.

Computer Simulation Models For Designing And Evaluating Alternative Population Planning Policies*

THOMAS H. NAYLOR

Duke University
Durham, North Carolina

World Population and the Problem of Economic Development

There is an increasing concern today for the effects of population growth and structure on the rate and structure of economic development—particularly in the case of the less developed countries. There is also concern for the reverse, yet mutually interdependent, effects of the present patterns of socioeconomic development on demographic processes. Spengler [1] has succinctly expressed the serious nature of the population problem by paraphrasing the words of John F. Kennedy:

Unless man halts population growth, population growth will halt man. As matters stand, two demographic processes are bound to bedevil man increasingly and, if they are not soon resolved, will

* This research was supported by National Science Foundation Grant GS-1926. We are indebted to Professor Robert Rehder of the University of North Carolina for a number of helpful comments.

greatly diminish whatever prospect he has of establishing a society both peaceful and great, both moral and aesthetic. These two processes are continuing population growth in a finite world and increasing concentration of people in progressively larger metropolitan centers. Progeny and space will have to be rationed much more effectively in the future than in the past.

Current anxiety with the population problem on the part of demographers, economists, and political leaders stems from the fact that there has recently been a marked shift upward in the rate of population growth.

and in many parts of the world to high levels for which there is no parallel in the past, and for which, therefore, the existing patterns of social and individual behavior may not provide an adequate response. Whether the problem lies in the possible adverse aggregative effects on per capita product, which is likely to be the case in the less developed countries; or in the differential impact on various components in the population, which is likely to be the case in more developed countries, a marked increase in the rate of population growth naturally leads one to ask whether the resources of knowledge and patterns of usual behavior are adequate. . . . Adequate response . . . means not merely the imaginative capacity to supply substenance for a larger population . . . it also means perception of feasible alternatives whose costs can be properly weighed and whose attainment is within reasonable grasp of the populations involved, rather than feats of derring-do by benevolent despots shaping docile populations at will.[2]

To consider a specific example, Gunnar Myrdal has painted a vivid picture of the urgency of the population crisis and the need for population control in South Asia in his recent book entitled *Asian Drama*. He points out that it is now commonly recognized that all countries in South Asia "have entered a critical phase of sharply accelerated population growth, and that the prospects for successful economic development are crucially related to population trends." [3] Myrdal refers to the situation in South Asia as "a veritable demographic revolution, the pace and dimensions of which are without precedent." Much of what Myrdal has said of Asia is also relevant to most of Latin America and many of the countries of Africa. Unfortunately, this demographic revolution was not foreseen by demographers until fairly recently because of their tendency to place all too much faith in simple extrapolations of earlier population trends.

Frequently, population planning efforts in the less developed countries have taken the form of hasty attempts to introduce western contraceptive practices into the cultures of Asia, Africa, and Latin America, which differ considerably from the culture of the western world. This approach to population planning has become the subject of increasing controversy in recent years. Current research buttressing present population planning efforts has often been limited to uncoordinated strategies to improve persuasion techniques and delivery system networks, to increase our knowledge of human reproduction and ways to control it, and to improve the collection of demographic data. Only lip service has been given to the fact that population planning involves an integrated, multi-disciplinary effort to disseminate the small family norm into the cultures of Asia, Latin America, and Africa. Also, the increased use of contraceptives is only one of the many tools available to the population policymaker.

Finally, the discussion of long-run goals and objectives is conspicuously absent from much of the literature on family and population planning.[4] Unless one specifies explicitly the goals of a particular population planning program, it is likely to be very difficult, if not impossible, to evaluate the program and propose suitable changes in its design.

In the light of the present population crisis, there appears to be a definite need for improved techniques for designing and evaluating alternative population planning policies. What is needed is a set of tools which will enable the policy maker to obtain answers to the following types of questions: What effect, if any, will a particular population planning policy have on population growth, per capita income, and other socioeconomic variables for a given country or region? Is a particular policy consistent with the national goals of the country in question? Given a set of politico-socioeconomic goals for a country, what kinds of population planning programs are likely to be "optimal," i.e., consistent with these goals?

Objectives

The Carolina Population Center Systems Analysis Program has recently begun an interdisciplinary research program whose objective is to develop a set of tools which can be used by population policy makers to obtain answers to questions of the type raised in the preceding paragraph.

What is proposed is the development of a collection of *computer simulation* models to (1) test the effects of alternative population planning policies on the politico-socioeconomic behavior of particular regions or countries and (2) design "optimal" population planning policies which are consistent with given social, political, and economic goals.

We shall define *simulation* as a numerical technique for conducting *experiments* with mathematical and logical models describing the behavior of a system on a digital computer over extended periods of time.[5] In our case, a demographic system includes biological, sociological, political, and economic elements. There are several key words in our definition of simulation which merit special attention.

First, the fact that simulation is a numerical technique implies that it is a technique of "last resort" to be used only when analytical techniques are not available for obtaining solutions of a given model. Being a technique of last resort by no means implies that simulation will find only limited usefulness in demography, for it is well-known that only a small number of problems in demography give rise to mathematical models for which standard analytical techniques exist for finding solutions. For most problems in demography, simulation may be the only technique available to the analyst.

Second, a computer simulation is an experiment. With the advent of the high-speed digital computer, social scientists can now perform controlled, laboratory-like experiments in a manner similar to that employed by physicists and other physical scientists, only using a mathematical model programmed into a computer rather than some other physical process such as a nuclear reactor. The only difference between a simulation experiment and a "real world" experiment is that with simulation the experiment is conducted with a model of the real system rather than with the real system itself. Since a simulation is an experiment, special consideration should be given to the problem of *experimental design*—a point which has been ignored all too often by demographers and other social scientists.[6]

Third, although a computer is not a necessary tool for carrying out a simulation experiment with a mathematical model of a demographic system, it certainly speeds up the process, eliminating computational drudgery, and reduces the probability of computational error. We anticipate that most of the models to be developed by the Carolina Population Center will be sufficiently complex to require a computer.

Fourth, we shall be concerned primarily with *dynamic* simulations, i.e., simulation experiments over extended periods of time.

Fifth, most simulation experiments with demographic models are stochastic simulations as opposed to purely deterministic simulations. Demographic models often include random variables over which policy makers can exercise little or no control. By including these random or stochastic variables in the model, a simulation experiment can be used to make inferences about the overall behavior of the system of interest based on the probability distributions of these random variables. Deterministic simulations are characterized by the absence of random error, i.e., all stochastic variables are suppressed.

An important indirect benefit of computer simulation model building is that in the process of formulating the model one necessarily learns a great deal about the system which is to be simulated, even if simulation experiments are never conducted with the model.* This is likely to be a very important benefit in developing models of family planning systems. The development of operational computer models of the type we

---

*Editor's note: An interesting approach to the simulation of certain economic dynamical systems which may have general relevance to the simulation of social systems is that of Jay W. Forrester (see his book *Industrial Dynamics,* MIT Press, 1961). Professor Forrester's work illustrates some of the points made in the present paper.

are proposing is likely to take many months, yet family planning program administrators need answers to their problems *now*. The knowledge gained about a particular population planning program while developing a model of it may very well lead to a more systematic approach to the solution of family planning problems prior to the completion of the computer model. For this reason, computer models of demographic systems offer the possibility not only of long-run benefits to population control policymakers, but also of immediate, short-run results.

Having defined exactly what we mean by simulation, and having indicated some of the reasons why we have turned to this tool as a means of designing and evaluating alternative population planning policies, we now return to the main point of this section of the paper.

Specifically, we plan to develop four different types of computer models and conduct simulation experiments with these models.

1. *Biological Models*—models for studying the determinants of human reproduction of the type proposed by Ridley and Sheps.[7-10]

2. *Population Models*—population projection models of the type developed by Orcutt [11] and models which attempt to "explain" changes in population growth such as those of Freedman,[12] and Schultz.[13,14]

3. *Economic Benefit Models*—projected income models [15] and investment models [16-19] which show the effects of population growth and population planning on economic variables, such as income and investment.

4. *Large-Scale Econometric Models*—models such as those of Brookings,[20] Wharton,[21] FRB-MIT,[22] and OBE [23] as well as models of less developed countries.[24-28]

After developing a set of models of the type outlined above for a given country or region which possesses a population "problem," our ultimate objective is to link these mutually interdependent models and thus create a closed model for the purpose of conducting policy simulation experiments.

Previous Work

The four types of models which were briefly described in the preceding section provide a convenient taxonomic device for summarizing the previous work in the population field which is relevant to the project proposed by the Carolina Population Center.

1. Biological Models

Previous efforts on the part of social scientists toward understanding the relationship of various factors to fertility and toward disentangling their influence have been concerned primarily with economic, social, and psychological variables and, to a great extent "have ignored the biological basis of human reproduction." [7] As Ridley and Sheps have pointed out, it is possible that "by overlooking the role of biological factors, social scientists have handicapped their efforts to understand the role of social and psychological factors." [7-10] As an example of this type of research, they have developed a simulation model to "investigate the quantitative effects on reproductive performance of changes in such factors as mortality, marriage patterns, use of contraceptives and their effectiveness, size of the family desired, fecundity and pregnancy wastage." Other studies in this field include the work of Hyrenius and Adolfsson [29] and Potter and Sakoda.[30]

2. Population Models

Projection Models. Projections of population and households have been a subject of principal concern of demographers for a long time. The path-breaking work of Orcutt, Greenberger, Korbel, and Rivlin [11] is now well-known to both economists and demographers alike. By constructing a microanalytic model of the United States household sector, Orcutt et al. demonstrated the potential of computer simulation as a tool for population projection and for testing the effects of changes in birth, death, marriage, and divorce rates on population growth. More recently the U.S. Bureau of the Census has become interested in population projection models. For example, Ackers [31] has developed a computer simulation program for the Census Bureau which has been used to make projections of marital status, to serve as an input to projections of households and families, and to evaluate marriage data from registration and from censuses and surveys. Among the other projection models which have been proposed by the Census Bureau are models to project mortality, fertility, family formation, and household formation.

Explanatory Models. Economists have shown a particular interest in a class of population models which we shall call explanatory. These models have as their objective the explanation of changes in population growth rates with particular emphasis placed on the factors which influence the birth rate. As an example of this type of model, Schultz [13,14] has formulated an econometric model which attempts to explain the frequency of births in a population in terms of three groups of factors that influence parents' desires for births: (1) the family size goal or number of *surviving* children that parents want; (2) the incidence of death, mainly among offspring, which necessitates an adjustment in birth rates to achieve any given family size goal; (3) the effect of uncertainty in the family formation process where births, deaths, and remarriage are unpredictable.[13] Schultz has tested his model with population data from Puerto Rico. Other studies which have been concerned with explaining population changes include Becker's [32] application of demand theory to human fertility and the work of Freedman [12] and Freedman and Coombs.[33]

3. Economic Benefit Models

By *economic benefit models* we mean models which attempt to measure quantitatively the "benefits" to society of *increases* or *decreases* in fertility. If one is able to ascertain the benefits of a particular population control program through the use of such a model, and if he knows something about the costs of the program, then he is in a position to say something about its cost-effectiveness. Horlocher [34] has written an excellent survey paper on economic benefit models in which he subdivides these models into two different categories—projected income models and investment models. Projected income models attempt to measure benefits in terms of projected per capita income with and without fertility reduction. Investment models attempt to estimate the effects in the future of a birth prevented now in terms of consumption need foregone, and the increase in government and private savings per capita.[34]

Projected Income Models. One of the first projected income models was Coale and Hoover's [15] econometric model of Indian economic growth, which attempted to measure the effects of reductions in fertility on per capita income. This type of model has been the subject of considerable controversy. For example, Myrdal has concluded that, in order to be useful, models of this type

> would have to contain many more parameters and account for many more interrelationships. They would have to be very much more complex in order to be logically consistent and correspond to reality. With the present dearth of empirical data, indulging in this type of preparatory macroanalysis does not seem to be a rewarding endeavor.[3]

The original Coale-Hoover model has been extended by Hoover and Perlman,[35] who applied it to Pakistan, and by Demeny,[36] who sought to answer the following question: "Assuming that the cost and effectiveness of a birth control program was known, what price would be worth paying for it and how would that price depend on the structural parameters of the economy?" [34]

Investment Models. The works of Enke,[16-19] Meier,[37] and Zaiden [38] treat a newborn child as though it were an investment and "the lifetime stream of consumption is deducted from the lifetime stream of production; the difference is then discounted back to the present by some rate of interest." [34] Horlocher, in the same work, reports on two efforts at a reconciliation of the projected income and investment approaches: the work of Julian Simon and Leonard Bower.

4. Large-Scale Econometric Models

Econometric models have now been constructed for many of the more highly developed countries including the United

States,[22,20,21,39,23] England,[40] France,[41] Holland,[42] and Japan,[28] as well as a few less developed countries, such as Italy,[24] India,[26] and Ecuador.[27] Of these models, the Brookings,[20] FRB-MIT,[22] OBE,[23] and Wharton [21] models of the economy of the United States, cited above, are probably the most sophisticated. In the four models of the United States economy, population is treated *exogenously*.* That is, no attempt is made to explain the two-way interaction which exists between population variables and the economy as a whole. In Shubik's [27] model of the economy of Ecuador, population is treated as an endogenous variable. However, he utilizes a somewhat naive model to explain changes in total population. Irma Adelman and Cynthia Morris [25] have recently proposed the inclusion of a number of social and political variables in large-scale econometric models.

A Critical Appraisal of Previous Work

Although considerable progress has been made in the last few years in conceptualizing population planning problems through use of mathematical models, there is still a great deal of work to be done before population models become operational tools of the policy maker. Therefore, it may prove useful to briefly summarize some of the more important shortcomings of the previous work in this field.

First, as Davis [4] has recently asserted, insufficient attention has been given by both policy makers and model builders to the objectives of population planning programs. Before one constructs a demographic policy simulation model and in turn conducts a simulation experiment, the objectives of the policy maker must be stated explicitly. Is he interested in controlling population variables for the sake of controlling population, or is he, in fact, more concerned with population variables because they affect certain socioeconomic variables? If the policy maker is interested in the effects of population planning on socioeconomic variables, then which socioeconomic vari-

---

* Strictly speaking, in the Brookings Model population is treated exogenously, but the proportions of the population in the labor force and marriages are treated as endogenous variables.

ables does he consider to be most important? In summary, as Myrdal has indicated, we must be more explicit in stating both our *quantitative* and *qualitative* goals of population planning.[3]

Second, causality seems to be uni-directional in most demographic models. Consider the following examples. Biological models are concerned with how mortality, marriage patterns, and the use of contraceptives affect the human reproductive process, but *not* with how human reproductive processes affect mortality, marriage patterns, and the use of contraceptives. Population models are concerned with how social, economic, and psychological variables affect population growth, but *not* with how population growth affects social, economic, and psychological variables. Economic benefit models are concerned with the effects of population planning (or the absence of it) on economic variables, but *not* with the effects of economic variables on population growth. Finally, large-scale econometric models can provide answers to questions about how population variables affect the economy as a whole, but *not* how the economy as a whole affects population growth. Yet we know very well that we live in a world in which causality may move in two directions rather than one and that events may take place simultaneously as well as sequentially.*

Third, the number of socio-economic and political variables included in existing demographic models is extremely limited. This is particularly true of what we have called explanatory models and economic benefit models. Although the large econometric models include a multiplicity of economic variables, social and political variables are conspicuously absent from most of these models.† Clearly, experience indicates that if we ignore social and political variables in attempting to implement a particular population control program in a given country, then the program is usually doomed to failure. It goes without saying that population planning is a complex problem, and to assume away many of the more important variables is equivalent to inviting disaster.

---

* **Editor's note:** In other words, feedback effects are important.

† Shubik [27] has included some political and social variables in his model of Ecuador.

Fourth, in view of the importance and complexity of population planning problems, it is simply unthinkable to entrust their solution to a single discipline such as public health or sociology, to mention only two possibilities. If we are going to develop operational models for designing and evaluating population control policies, we must first develop an integrated theory of population dynamics which draws heavily on the disciplines of economics, sociology, political science, psychology, and biological science. By its very nature, population planning is an interdisciplinary field.

Research Strategies

Given the objectives of the Carolina Population Center Systems Analysis Program, three possible research strategies appear to be worthy of consideration: (1) a computer game; (2) micro-analytic models; (3) macro-analytic models.

1. A Computer Game*

Gaming refers to a special type of simulation in which human participants act as decision makers within the framework of the system being simulated.

> A [simulation] game is a contrived situation which imbeds players in a simulated . . . environment, where they must make [policy] decisions from time to time, and their choices at one time generally affect the environmental conditions under which the subsequent decision must be made. Further, the interaction between decision and environment is determined by a refereeing process which is not open to argument from the players.[43]

What is envisaged is a computer game based on a socio-politico-economic model of a hypothetical underdeveloped country such as India. Although the parameters of the model would be contrived, every effort would be made to capture the essence of the economic, political, and social processes of

---

** **Editor's note:** This is not to be confused with the subject of "game theory."*

a prototype less developed country in the same manner in which business management games such as the Carnegie Tech Game [44] have captured the essence of industrial processes. The players or policy makers would have at their disposal numerous population control policies. A particular policy could be read into the computer; the computer would then simulate the long-run effects of the particular policy on population growth as well as other socioeconomic and political variables. The player could then experiment with a number of different policies, each time observing the simulated effects of the policies. Alternatively, the player could be asked to specify his population control objectives and then devise a set of policies which would give rise to simulated results that are consistent with these objectives.

To the extent that the game was "realistic," it would have enormous potential as a teaching device and a planning tool. The experience alone gained by building a hypothetical model of this type, before turning to an empirical model aimed at a particular population program or a particular country, would be considerable.

In other words, the creation of a gaming model of a hypothetical environment might prove to be a very useful transitional step prior to building an *operational* model based on actual data.

2. Micro-Analytic Models

Micro-analytic models focus on individual decision processes rather than highly aggregated processes. A micro-analytic model might be useful in evaluating the behavior of *individuals* in response to a particular family planning program. Computer models for studying the determinants of human reproductive processes, as well as cost-effectiveness models of specific population control programs, would be micro-analytic in nature. Orcutt's [11] population projection model and Schultz's [13,14] explanatory models are examples of decision oriented, micro-analytic models. Most of the biological models are also micro-analytic in nature.

One approach to the development of a total population system model of a less developed country would be to develop a set of micro-analytic models for different social and economic sectors of the country and link these sectors in such a way as to close the system. With this modular approach one could perform experiments with a particular sector or sectors while holding everything else constant, or experiments could be conducted with the model as a whole.

Although the modular approach to model building has been espoused by many social scientists, with the possible exception of the Brookings Model of the United States there are not very many examples of its actual use. Unfortunately, relatively little is known about the problems of linking separate micro-analytic models. One of the objectives of this proposed study is to explore the problems of linking biological and socio-economic models in an effort to construct a total population system model.

As we have previously mentioned in outlining the objectives of this study, a number of short-run benefits may accrue to family planning administrators from the very process of developing a computer game or a micro-analytic model. That is, the kind of questions which must be raised and answered in the process of formulating a computer model of a family planning program and service delivery system are likely to prove to be invaluable to family planning administrators long before the model itself is completed and operational. For the family planning administrator in an underdeveloped country who cannot wait six months for a model to be completed before making a crucial decision, the possibility of obtaining a conceptual framework for attacking his immediate problems may prove to be an important advantage of computer model building efforts.

Therefore, a major aim of this project will be to search for *short-run solutions* to existing problems while simultaneously developing computer models which will become operational for family planning administrators within one or two years. Unless some attention is given to the very pressing problems of the present, the magnitude of population problems may have increased to such an extent by the time we complete our computer models that they will have been rendered completely useless.

3. Macro-Analytic Models

Macro-analytic models for example, large-scale econometric types, are highly aggregative in nature and are more concerned with the overall behavior of a complex socioeconomic system rather than with the individual decision processes.

We propose to build a large-scale econometric growth model for a given less developed country which is closed with respect to population. That is, population will be determined endogenously within the model and will be partly influenced by the socioeconomic variables of the model rather than being treated exogenously as it has been treated in existing models. As an initial step in this direction, we plan to append a population projection model to a modified version of the Wharton Model in order to gain experience with this approach before trying it with a less developed country where data are in extremely short supply. Ultimately, we would hope to build at least three similar models for less developed countries, including a Latin American, an African, and a Southeast Asian country.

Methodological Problems

It is clear that there are numerous serious methodological problems ahead of us on the course on which we are about to embark. We shall now summarize some of these and point to possible solutions if they are known.

First, although we realize that data problems are likely to be acute in many instances, we do not feel that these alone are sufficient to make computer simulation impractical as a policy-making tool. Even if the data for a particular variable are incomplete, we may still be able to treat it as a random variable and approximate its probability distribution or estimate its time trend and show the effects of this variable on the system being simulated. We may find that the system is not sensitive to the variable, in which case it can be eliminated. On the other hand, the system may be quite sensitive to changes in this variable, in which case we can run several simulations based on the most likely values of the missing variable. In other words, all is not lost if data are unavailable for a particular variable.

Second, once we have formulated a given model we are then faced with the problem of estimating the parameters of the model based on the available sample data. However, a problem arises. Computer models are dynamic models and unfortunately, relatively little is known about the statistical performance of dynamic models whose parameters have been estimated using standard statistical techniques. To illustrate this problem, we shall consider econometric models consisting of sets of simultaneous, stochastic difference equations. Although the static properties of estimators such as ordinary least squares, two-stage least squares, and full information maximum-likelihood methods are well-known, we have no assurance whatsoever that a model whose parameters have been estimated by one of these methods will yield *valid* dynamic, closed-loop simulations. That is, it is quite possible for a model which has been estimated by one of the aforementioned techniques to yield simulations which in no sense resemble the behavior of the system which they were designed to emulate. What is needed is an estimation technique which uses as its criteria of goodness-of-fit, "How well does the model simulate?" rather than, "How well does the model confirm past historical results?"

Third, computer models which appear to have been estimated properly may produce simulation results which are nonsensical. That is, the simulations may explode, turn negative, or lead to results which are in complete conflict with reality. We must learn more about the mathematical properties of our models with the hope of devising techniques which will enable us to spot these problems with our models analytically before running simulations with them. There appears to be a definite need to combine the approaches of the econometrician and the systems analyst in formulating models of complex socioeconomic systems. To the econometrician a model of a complex socioeconomic system consists of a system of simultaneous difference equations. To the systems analyst a demographic model consists of a set of mathematical inequalities which reflect the various conditional statements, logical branchings, and complex feedback mechanisms that depict the economy as a dynamic, self-regulating system. Finally, if we are going to build realistic simulation models which accurately reflect the underlying decision processes of a complex socioeconomic system, we are going to find it necessary to draw heavily on a multiplicity of bio-behavioral disciplines.

Fourth, as we have mentioned before, careful attention should be devoted to the problems of experimental design and data analysis in implementing our computer simulation experiments. Here, we shall draw heavily on the previous work of the Econometric System Simulation Program at Duke University.[6][45-48]

REFERENCES

1. Joseph J. Spengler, "The Economist and the Population Question." *American Economic Review, 56,* March 1966, pp. 1-24.
2. Simon Kuznets, *Demographic Aspects of Economic Growth.* United Nations Population Conference, Belgrade, Yugoslavia, August 20-September 10, 1965.
3. Gunnar Myrdal, *Asian Drama: An Inquiry Into the Poverty of Nations* (New York: Twentieth Century Fund, 1968) .
4. Kigsley Davis, "Population Policy: Will Current Programs Succeed?" *Science, 168,* November 10, 1967, pp. 730-739.
5. Thomas H. Naylor, Joseph L. Balintfy, Donald S. Burdick and Kong Chu, *Computer Simulation Techniques* (New York, John Wiley & Sons, 1966) .
6. Thomas H. Naylor, Donald S. Burdick and W. Earl Sasser, "Computer Simulation Experiments with Economic Systems: The Problem of Experimental Design." *Journal of the American Statistical Association, 62,* December 1967, pp. 1315-1337.
7. J. C. Ridley and M. C. Sheps, "An Analytic Simulation Model of Human Reproduction with Demographic and Biological Components." *Population Studies, 19,* March 1966, pp. 297-310.
8. J. C. Ridley *et al.,* "The Effects of Changing Mortality on Natality." *Milbank Memorial Fund Quarterly, 45,* January 1967, pp. 77-96.

9. Mindel C. Sheps, "Applications of Probability Models to the Study of Patterns of Human Reproduction." *Public Health and Population Change,* N. C. Sheps and J. C. Ridley, eds. (Pittsburgh: University of Pittsburgh Press, 1965).

10. Mindel C. Sheps, "Contributions of Natality Models to Program Planning and Evaluation." *Demography, 3,* 1966, pp. 445-449.

11. Guy H. Orcutt, Martin Greenberger, John Korbel and Alice M. Rivlin, *Micro-Analysis of Socioeconomic Systems: A Simulation Study* (New York: Harper & Row, 1961).

12. Deborah S. Freedman, "The Relation of Economic Status to Fertility." *American Economic Review, 53,* June 1963.

13. T. Paul Schultz, "A Family Planning Hypothesis: Some Empirical Evidence from Puerto Rico." The Rand Corporation, RM-5405= RC/AID, December 1967.

14. T. Paul Schultz, "An Economic Model of the Family Planning and Fertility." The Rand Corporation, P-3862-1, July 1968.

15. A. J. Coale and E. M. Hoover, *Population Growth and Economic Development in Low-Income Countries* (Princeton: Princeton University Press, 1958).

16. S. Enke, "Speculations on Population Growth and Economic Development." *Quarterly Journal of Economics, 71,* February 1957, pp. 19-35.

17. S. Enke, "The Gains to India from Population Control: Some Money Measures and Incentive Schemes." *Review of Economics and Statistics, 42,* May 1960, pp. 175-180.

18. S. Enke, "The Economics of Government Payments to Limit Population." *Economic Development and Cultural Change, 8,* July 1960.

19. S. Enke, "The Economic Aspects of Slowing Population Growth." *Economic Journal, 86,* March 1966.

20. J. S. Duesenberry *et al., The Brookings Quarterly Econometric Model of the United States* (Chicago: Rand McNally, 1965).

21. Michael K. Evans and Lawrence R. Klein, *The Warton Econometric Forecasting Model* (Philadelphia: University of Pennsylvania, 1967).

22. Frank DeLeeuw and Edward Gramlich, "The Federal Reserve-M.I.T. Econometric Model." *Federal Reserve Bulletin,* January 1968, pp. 1-40.

23. M. Liebenberg *et al.,* "A Quarterly Econometric Model of the United States." *Survey of Current Business, 46,* May 1966, pp. 13-39.

24. G. Ackley, *Un Modelle Econometrico dello Sviluppo Italiano nel Dopoguerra* (Roma: Giuffre Editore, 1963).

25. Irma Adelman and Cynthia Taft Morris, "An Econometric Model of Socio-Economic and Political Change in Underdeveloped Countries." Unpublished manuscript, Northwestern University, December 1967.

26. N. V. A. Narasimham, *A Short-Term Planning Model of India* (Amsterdam: North Holland, 1956).

27. Martin Shubik, "An Aggregative Socio-Economic Simulation of a Latin American Country." Cowles Commission Discussion Paper No. 203, Yale University, August 1, 1967.

28. H. Ueno and S. Kinoshita, "A Simulation Experiment for Growth with a Long-Term Model of Japan." *International Economic Review,* 9, February 1968.

29. H. Hyrenius and I. Adolfsson, "A Fertility Simulation Model." Demographic Institute, University of Göteborg, Sweden, 1964.

30. R. G. Potter and J. M. Sakoda, "A Computer Model of Family Building Based on Expected Values." *Demography, 3,* 1966, pp. 450-461.

31. Donald S. Ackers, "On Measuring the Marriage Squeeze." *Demography, 4,* 1967, pp. 907-924.

32. Gary S. Becker, "An Economic Analysis of Fertility." *Demographic and Economic Change in Developed Countries.* National Bureau of Economic Research (Princeton: Princeton University Press, 1960).

33. Ronald Freedman and Lologene Coombs, "Economic Considerations and Family Growth Decision." *Population Studies, 20,* November 1966.

34. David E. Horlocher, "Measuring the Economic Benefits of Population Control: A Critical Review of the Literature." Working Paper No. 2, Penn. State-U.S. AID Population Control Project, May 1968.

35. Edgar M. Hoover and Mark Perlman, "Measuring the Effects of Population Control on Economic Development: Pakistan as a Case Study." *Pakistan Development Review, 6,* Winter 1966.

36. Paul Demeny, "Investment Allocation and Population Growth." *Demography, 2,* 1965, pp. 203-233.

37. Richard L. Meier, *Modern Science and the Human Fertility Problem* (New York: John Wiley & Sons, 1959).

38. G. Zaiden, "Benefits and Costs of Population Control with Special Reference to U.A.R." Unpublished Ph.D. Dissertation, Harvard University, 1967.

39. Gary Fromm and Paul Taubmann, *Policy Simulations with an Econometric Model* (Washington, D.C.: The Brookings Institution, 1968).

40. L. Klein *et al., An Econometric Model of the United Kingdom* (Oxford: Basil Blackwell, 1961).

41. Michael K. Evans, "A Short-Term Forecasting Model of the French Economy." Unpublished manuscript, University of Pennsylvania.

42. Central Planning Bureau, *Central Economic Plan 1961.* The Hague, August 1961.

43. Proceedings of the Conference on Business Games as Teaching Devices. Tulane University, April 26-28, 1967, p. 7.

44. Kalman J. Cohen *et al., The Carnegie Tech Management Game* (Homewood, Illinois: Richard D. Irwin, 1964).

45. Thomas H. Naylor, William H. Wallace and W. Earl Sasser, "A Computer Simulation Model of the Textile Industry." *Journal of the American Statistical Association, 42,* December 1967, pp. 1338-1364.

46. Thomas H. Naylor, Kenneth Wertz and Thomas Wonnacott, "Methods for Analyzing Data from Computer Simulation Experiments." *Communications of the ACM, 10,* November 1967, pp. 703-710.

47. Thomas H. Naylor, Kenneth Wertz and Thomas H. Wonnacott, "Some Methods for Evaluating the Effects of Economic Policies Using Simulation Experiments." *Review of the International Statistical Institute,* February 1968.
48. Thomas H. Naylor, Kenneth Wertz and Thomas H. Wonnacott, "Spectral Analysis of Data Generated by Simulation Experiments with Econometric Models." *Econometrics, 37,* April 1969.

II
Mathematical Aspects

Mathematical and Computer Models of Large Systems

JOHN G. KEMENY

Dartmouth College
Hanover, New Hampshire

I would like to talk about what I feel are some current shortcomings of mathematics in the modeling of very large systems, and to present some rather hazy ideas of how I think computers may come to our rescue.

Naturally, any mathematician starts out with the prejudice that the only worthwhile model, the only one we should pay serious attention to, is the formal mathematical model. I held that prejudice for many years myself and it was only in trying to dabble a bit in the applications of mathematical models to the social sciences that I came to the conclusion that the short-comings, at least in the short run, far outweigh the advantages. I suppose I still feel that, in the long run, as a science becomes more mature, as it grows up, as it becomes more systematic, it should become fully mathematical and should follow in the footsteps of the classical and successful sciences. However anyone who hangs around students in academic life nowadays, comes to the conclusion that one can't live just for the long run. One must make decisions for the short run, and the problems of urban society cannot wait until mathematics grows up to the point where it can come up with beautiful models of the problems of urban society.

In trying to pinpoint why mathematics is as poor as it is it seems to me mathematics is very powerful at the two extremes of the scale.

At one extreme there are a small number of parts to a system of fairly simple connections. Mathematics can be extremely good in exhausting all the possibilities by getting a complete solution of the set of equations, or whatever the case may be.

At the other extreme, we have a marvelous trick when the number of particles gets to be of the order of magnitude of some very high power of ten. If they are sufficiently homogeneous, the mathematician pretends that there are infinitely many of them. He assumes certain properties of continuity, some sort of smoothness of properties that enable him to pass to the limit, and then the whole power of classical analysis comes into the play. Most of the history of modern science is connected with the tremendous benefits reaped from this approach.

Unfortunately, most of the large systems we are talking about when we are interested in problems of society fall into a range that is intermediate to these. The problems are much too large to get explicit solutions for them, and yet, the number of parts is not large enough, nor are the parts homogeneous enough, to be able to pass to the limit.

Many mathematicians, myself included, have predicted that entirely new branches of mathematics will have to be invented someday for the solution of problems of society, just as a new mathematics was necessary before significant progress could be made in physics. I still believe this, but I think I'm now much more cautious on how long it will take to come about. As I started out saying, we cannot wait until this great revolution comes about in mathematics.

In the meantime, I would like to point out at least three areas where mathematics is at the moment having very peculiar effects.

The great prejudice of practitioners of mathematics is for

the continuous. If I wanted to caricature what happens in a typical application of mathematics—and I'm afraid the caricature is a little too close for comfort—it would go as follows.

You start with a problem that's clearly discrete. It has a finite number of parts, with a finite number of connections. You're dealing with a finite amount of time, very often in discrete stages. However, you are used to working with continuous models and, therefore, you pretend that you have a continuous model and you come up with a very lovely partial differential equation. At this point, the model-former is very happy, and he hands the problem over to his colleague who is an expert on solving partial differential equations.

His colleague takes one look at the equation and says, of course, "This cannot possibly be solved in closed form." Therefore, he uses his great tool, which is to approximate the continuous equation by a nice discrete model, and then goes and solves it on a computer. Well, if someone happens to be there during the entire process, there are two things that are disturbing. One of them is that the discrete model we end up with has significant differences from the discrete model problem we started with, and second, it is not entirely clear in many of these applications how we were helped by the fact that we went through the stage of forming continuous equations.

The second thing I would like to criticize in the applications of mathematics—and incidentally, criticism here is criticism in the best form because it started out as self-criticism—is that mathematicians have terrible prejudices about coming out with exact or optimal solutions. Since I have had a good deal of hands-on experience with computers and have been finding out the difference between doing things in principle and doing them in practice, I have discovered that the difference between an optimal solution and an outstandingly good solution can be one or two orders of magnitude in difficulty and in time. I am beginning to feel that mathematics at the moment is exerting a pernicious influence in that it is somehow conditioning us to believe that a solution to a problem is not really good unless we can prove that we have achieved the ultimate in it. We must have an exact solution (whatever that may mean, because, of course, before it becomes practical, we are going to come up with a good approximation to it and apply that). We must come up with an optimal solution and we must be able to prove that the solution is optimal.

I guess I am beginning to be influenced, as I said before, by the impatience of our students on the campus. But, I don't think we can wait until the theoretical tools get to the point where we come up with best solutions. In many problems one hopes that one might with a reasonable exertion of effort come up with an order of magnitude improvement; one should not wait until one can substitute for this an exact or optimal solution.

The third and final prejudice introduced by mathematical models is really the fault of the educational system rather than of mathematics itself. Certain traditions have come about as to what mathematics is really important for applied mathematicians, such as physicists, biologists and sociologists, and I'm afraid that these traditions simply do not correspond to the needs of the practitioners.

For example, in the typical physics curriculum one finds—and again I'm caricaturing it—that you start with calculus and then you take more calculus and then you take more advanced calculus and, if you have time, you take some extremely advanced calculus; so that you are going to be very well prepared for the future needs of your profession. In an age where the most difficult problems of physical science deal with random processes, with stochastic processes, and the interpretations and treatments of these, I have seen any number of recommended curricula where they never found room for a single course in probability theory.

The reason I feel this is very dangerous is that classical analysis has had so much more success in application to science than have the other branches of mathematics that the scientists who come up with models don't know the other branches of mathematics. If you will respond to this that in part the fault lies with the mathematicians, I think you are quite right; we have to accept a significant share of the blame.

I certainly am not going to come out and say that I feel that computers are going to present a utopia; very far from it. However, I would like to point to some features of computer models that seem to me encouraging.

First, let's make clear the way I use the word model. At some stage a model may have been some sort of small physical system that paralleled the action of a large system; at some later stage it may have been a verbal description of that system, and at a later—and hopefully more advanced—stage, it may have consisted of mathematical equations that somehow described the behavior of the system. I think we have to accept certain types of computer codes as models; this is not even an extension of the way the word model has been used traditionally. However, using computer codes has one very obvious advantage: namely, that they are written in terms of a very simple language—a language that to me may be the new universal language of scientists. Learning one of the standard computer languages, whether it will be ALGOL, BASIC, or FORTRAN, is vastly easier than, let us say, learning partial differential equations. Equally important, it is much more neutral as to types of mathematical approaches than is any single branch of mathematics. For this reason, I think this is an intrinsically worthwhile approach to the modeling of scientific problems. Computer codes are a great deal easier to modify than most systems of mathematical equations, and they are very well adaptable to the introduction of random processes. The simulation of random processes on computers is quite well understood, and it is much easier to introduce a random process into a computer code than to introduce it into something that is described by a partial differential equation.

But, above all, I like the approach through computer models, because once you get into an argument you can argue on a much smaller scale. 1 sometimes find arguments on a very broad and general scale too frustrating; the chances for agreement are almost nil. I like arguments on a level where you may have a large number of boxes in your flow-diagram and you start to argue as to whether the effect of this particular box which represents one factor in the total social system is correctly incorporated into the system. One is much more likely to get agreement at this level, and I think that the kind of discussion one gets into and the kind of test one would perform to find out how to correct and improve the model is much more useful and fruitful.

I know what the standard answer is to saying that one ought to have computer models. The standard answer is yes, but it takes too much computing time. They are too complex; it is too hard to get practical results out of them. I would, therefore, like to propose four points of what one can do with computer models which might lead to a significant improvement in the applicability of computer models to scientific problems.

The first one simply repeats something I said earlier, that one should not try to find exact or optimal solutions to problems, because it may cost one or two orders of magnitude more time and money to come up with an optimal solution to a problem than to come up with something that is simply very good. I think we should be less purists and more willing to be pragmatic in a transitional stage in the social sciences, and start working on methods that will give significant improvements rather than on methods that are known to be optimal.

Second, it seems to me that we are too much conditioned by the necessity of applying very complicated and powerful tests of significance to our models. I have seen computer models done by really excellent people, where most of the time was spent in checking the significance of what they were doing. At least once I had quite a good deal of success by proposing a childish approach to the same problem. They were worried about the question of how much effect the random variables had on the model. They were in the process of going through enormously complicated statistical analyses to try to estimate the effect of using a different set of random numbers. I suggested that instead they take the original problem and run it three times (or ten times if they had enough computer power and time available), using different sets of random numbers, and simply look at the results to see if they differed a very great deal. The answer in that particular model was that they ran it three times, and the results differed by such an enormous amount that it was perfectly clear that it was ridiculous to try to apply very detailed statistical tests.

Similar kinds of tests can be applied in the question of the meaningfulness of parameters. One always worries about the fact that in social problems measurement is very difficult, that there is likely to be a very large error in measurement of parameters and the data you are starting with. Well, a very simple-minded approach to this is to take a guess as to how far off your data is likely to be and just run several examples of the same model, arbitrarily modifying data by about the right order of magnitude, and see whether the results you come out with are qualitatively different or not. If they are, as is very often the case, then your model is useless. You can then worry about getting better measurements or getting models that are not quite so sensitive to differences in the parameters.

The third feature I would like to mention is the one I feel most strongly about, and perhaps I may get a chance to answer the question of how one introduces dignity into computer models. My thinking on computers has been changed rather drastically in the last four years by our experiences with a quite large time-sharing computing system at Dartmouth. If I may put in one commercial for Dartmouth, we have developed the system ourselves in the last four years; we have trained eighty-five per cent of all our students in how to write a nontrivial computer program, and I think this has had some rather dramatic effects on liberal arts education. What I really wanted to talk about is not the marvelous things we did at Dartmouth, but the things we didn't think about at all, that came about purely by accident.

We completely underestimated the effect of man-machine interaction. Unfortunately, today, most people who have computer experience have little or no experience in man-machine interaction. I am simply going to state that everyone who has ever come in contact with this to any degree has been astounded by its effect. Therefore, my fourth point is that we should stop thinking about building complicated models, which we work on for a long time and then give to a computer to let it work on for a hundred hours. I'm totally convinced that this is basically the wrong way to approach computer models. A computer model should consist of an interaction between a human being and a machine; the machine carries out the millions or billions of steps of computations that are needed, but it must have frequent and significant interaction with a human being who is an expert in that particular field.

There are a number of cases where the programming time can be significantly reduced by man-machine interaction. But, much more important, there are a number of activities that human beings are extremely good at which they are perfectly miserable at in explaining to other people. If they cannot explain it to other people they are much worse in explaining it to computers. I think the fact that it has been demonstrated that some of these things can be explained to computers is irrelevant. If it's going to take you much longer to explain it to a computer, and the computer will do it much worse than you do yourself, why force the computer to do it?* The answer today is that most computing systems are run as batch-processing systems; you simply don't have a chance to step in and make a crucial decision. But this is a purely temporary stage in the development of computers, and I think that research scientists should think more and more about interactive models, so that the scientist can provide the intuition, the value judgment, and the decision on how to proceed.

If I may answer Mr. Walter's very well-chosen challenge, man-machine interaction can provide the dignity in the model. Consider an extreme and ludicrous example: It's terribly easy to come up with a computer model where you extrapolate some highly simplified model as to how New York City is going to

---

*Editor's note: Norbert Wiener also emphasized this point using machine translation as an example. In this task the computer is useful, but only up to a point. It can do a large amount of tedious routine work with extreme rapidity; however the final stage of the translation must be performed by a human who makes the final choice of words and sentence structure. Without such final editing the translation would have very limited utility in most cases. This simply illustrates that humans and computers are good at doing different types of tasks. The only practical thing one can do is employ both of them together in a mutually complementary manner as Kemeny suggests here.

It is interesting that in 1963 Wiener felt that computers would not evolve to the point where they could do the complete translation task without human intervention in the foreseeable future.

develop over the next 200 years, and because of some factor you have overlooked, the model does not note that, at the end of year seventeen, there would have been a revolt in the city with large-scale slaughter if things really went that way. If the modeler could once a year look at how things were going in his model, he could step in and say, "No, no, no! At this point something completely different would happen."

My final point is that if it is so very costly to come up with these interactive models, I would like to propose certain dual-purpose computer systems. I think there is too much of a separation between systems that are designed to perform services, for which there is often a great deal of money available, and systems that are designed for theoretical development, or data gathering, or for learning how to do things. I think there could be a very happy marriage of needs and opportunities if one starts to think of designing new systems that are dual-purpose.

I'll pick one out of a hundred possible examples; one that for personal reasons I am very much interested in. Suppose New York City wanted to get into the act of doing something really intelligent in helping people get jobs. Instead of having 3,000 employment agencies, each of which has a small fraction of the available jobs listed, each of which has many clerks not particularly trained for the job who have no idea about the very difficult optimization problems of bringing together human beings with jobs, one could design a large time-sharing computing system for New York City which could operate with the same 3,000 employment agencies but where the agents would simply serve as an interface between the candidates and the machine. The machine could correlate the needs of thousands of individuals with available jobs, could hopefully do, not an optimal, but a decent job in matching up human beings with jobs, where a person from either a very good neighborhood or from a very poor neighborhood would have access to the same list of jobs, and would be judged impartially as to his qualifications. While you are doing all these wonderful things, if someone just thought about it and the system was cleverly designed, you could collect an immense amount of information on a real-time basis, with other people doing most of the work, as to what the major problems are at the moment, what the sources of unemployment are, where there may be a major mismatch between the needs of industry and the labor force, in what parts of the city they exist, where a change of the transportation system would help, where new education is needed, etc. We could, for the first time, reach the point where we at least have a chance, without bankrupting the whole system, of getting the kind of detailed, up-to-date data which would make intelligent, rational planning possible. Whether this will actually come about or not is, of course, beyond my competence to guess.

Cybernetics and the Management of Large Computer Systems

EARL C. JOSEPH

UNIVAC Division of Sperry Rand Corporation
St. Paul, Minnesota

Introduction

In an age when revolutionary change is the name of the game, it will come as a welcome surprise to most computer users to find that their next generation digital computers will evolve from today's systems. However, this will not be true of the impact of these computers—for their continued proliferation toward mass applications will affect and revolutionize society in a fashion which will stagger man's imagination. With the impact of computers exploding upon us on a worldwide scale, before the man in the street can react, we are at the threshold of a new revolution. This realization by society and the computer industry alike is in itself bringing about change.

The trend riding the wave of the future is a greater sensitivity by computer designers to the major problems of the computers impact, the problems of programming, and the management of large systems. A new revolution is occurring which overshadows the evolving next generation general-purpose computer; it is a new awareness that it is necessary to solve the special problems in every field.

I want to turn your thoughts to some of the problems in the management of large systems, primarily those problems that concern the digital computer and its use, and to take you somewhere into the future by showing you some problem solution trends. Emphasis will be on future digital computer systems rather than on projections of computer applications. First, I will draw some imperfect analogies of computer systems with biological systems. The reason for these analogies is not so we can simulate them, but because we have a level of understanding about such analogies—in some cases, a level greater than our understanding of our computer systems. Next, I will offer a view of the real world of large computer systems, by looking at the not-so-rosy side of computers—at some of the problems we are encountering today in the use of computers. I will examine the primary problems of large computer systems, of their management and control, communication, and programming to detect their impact on future system design. Finally, I will project a view of the rosy side of the future by previewing the vast changes that can be expected in the coming decade and some trends which solve today's problems. I will touch upon both hardware and software computer trends which capitalize on the wonders and solve the blunders of the past.

We are deep within the computer [1]/ communications [2] revolution, racing forward into the cybernetics revolution, but the man on the street scarcely knows of its existence. So far, most of the fruits of the computer revolution seldom enter directly into our daily lives, they touch us only indirectly. Who knows, for example, that without computers the many life-giving kidney and heart transplants would not be a reality, or that jets are here today because aircraft designers use computers, or that without the computer, many of us would not be paid promptly. The new president of IFIP, at the recent IFIP Congress '68 in Edinburgh, Scotland, described this state of affairs rather eloquently, although he gave it a slight political overture: "There have been three great discoveries in the last quarter of the current century: 1) The explosion of the first atomic bomb—which horrified everybody; 2) The first Russian in space—which everyone applauded; and 3) The computer revolution—which hardly anyone knows of. . . ."

There is evidence based on today's facts and established trends that next-generation computers will/could:

- Make engineers, scientists, managers, businessmen, politicians, and all of society more creative
- Its power will become easily and cheaply available to a great many users (the man on the street) through the use of remote terminals operating in a time-shared mode
- Sharply reduce the cost of decision making which will lead to better management
- Relieve man of many more tasks by performing increasingly more control operations over man, his machinery, and society
- Replace expensive laboratories and remove risks to society by allowing large-scale experiments to be simulated
- Bring back individualism rather than leading society toward dehumanization.

A similar amount of evidence exists which points to the negative side.

The computer is not automatically the maker of the cybernetics era, but it is a necessary tool in such an era. It is the key which unlocks the gates to the era, the means and the technology which permits it to happen. But the tool must be applied before we can enter the cybernetics era. Today we are rushing in that direction, but we have a tremendously long path ahead before we can hope to arrive. In this decade, however, we passed a major milestone. In the 1950's it was the computer manufacturer who was recommending computers and pushing them into exotic new areas; this is not so today. In this decade, it is the user who is pushing the computer manufacturer to supply newer, faster, and more powerful computers for new and far-out applications, some of which consume the imagination to contemplate.

Planning for the Future[3]

There are a number of reasons for looking to the future, to predict five, ten, thirty years and more into the future, especially for the computer industry—the manufacturers of computers. First of all, the computer heralds the future—it is used to plan for the future and to implement the future. Its role for the future is on a steep increase. More important, the computer is applied first, before other technology, in this role. In order for the computer of the future to meet the need, the computer manufacturer is forced to look at the trends in the application areas of computers, as well as the trends in computer technology. This is necessary so that future computer systems can be designed and built to meet the predicted needs and within the limits of technology. Without such predictions of the future, Murphy's law applies: "Any decision made without some knowledge will be the wrong decision." Second, part of the future is what we make it; its coming can be accelerated or impeded simply by the amount of money and effort we apply to bring it about. Further, we must predict that major part of the future over which we have no control in order to be ready for it when it does come about. We do not want to be in the position of continually trying to catch up with the future, for if this is the case, we will always be behind. We would rather leapfrog into the future through our predictions and planning.

If we can first know whither our computer systems are tending and future requirements for them, then we can better judge how and what to design to get them there. Our principal task, then, seems to be to find the trend and predict the future before we, as designers, initiate next-generation computer design efforts.

The rash of many programming problems and the creation of many new systems, both hardware and software, during the fifties and early sixties, attest to the vigorous undirected efforts which have apparently floundered, leading them to early obsolescence. Few predictions can be proclaimed as satisfactory unless the reason given is one based on solving the observable problems being experienced today with today's systems. The point is that, since passing the halfway point in this decade, system designers have slowly and painfully developed an awareness that programming costs the computer user more than the hardware. The designer needs, however, to go beyond this consideration alone by also predicting the future problems which will be encountered with his proposed solutions in order to allow his ideas to evolve. That is, those designs which are not born of experience are risky. Yet we must also leapfrog into the future by looking beyond today's problems.

The modern practice of technological forecasting is scientific in intention if not always so in execution. The discipline of programming seems at last to be growing up and promises to attain the status of a science, or at the very least, systems designers of next-generation systems are now able to recognize its latest fads. Since this is really the case, and government statistics show that it is, designers are beginning to realize that they need to build many new features into the hardware to ease the problems of programming. Designers are now forecasting solutions to the problem of programming.

To this end, computer designers have at last begun to emancipate themselves from trading-off hardware features against other hardware features alone by adding the major new trade-off area of programming as well as economic considerations to their design equations. They are not finding this an easy task, for the programmer is not much help. This is apparent when the programmer is asked what is needed; since the programmer seldom knows what designs fall within the realm of possibility, the programmer's answer is always couched in the constraints of the computer system with which he is most familiar. Computer designers are faced with an obvious dilemma; programmers can offer little help in suggesting ways to design future computers to ease the problem of programming, and designers, in the past, have not been sufficiently schooled in programming.

What then is the answer to the designers' question: What is needed? The obvious answer to this question—and fortunately the trend which is beginning to materialize—is for the computer designers to become schooled in the art and science of modern programming. And this is happening today as more and more designers are forced into programming in order to design and build adequate computer systems. That is, as designers apply computers to their everyday design tasks, they are brought face-to-face with the task of programming and they encounter the programming problems first hand. If the designers are fortunate enough to program in the environment of a typical computation center, using typical system programs (such as operating system, executive, and high level language programs), and interacting with other applications, the programming problems they face are typical and representative. When this is the case the result can be dramatic, as next-generation computer systems will demonstrate.

Analogies — Computers with Biological Systems

The subject of biological analogies of computers is not only fascinating but timely; it deals with placing present and future technology in a known or familiar mold. Generally speaking, mankind is entering this last third of the century with an enormous potential, which is largely predictable, for using technology to serve useful ends—especially in the areas of great social needs. Why not then draw analogies between computers in these areas and those in biological systems? The development of computer technology is undergoing an unusual acceleration, one of the greatest of all times. According to the Commerce Department, the computer industry in the last decade became one of the all-time growth champions by increasing its use by more than 500 percent over the previous decade. This growth is unusual in the sense that the change has been qualitative as well as quantitative. There is no longer any doubt that this technology has uplifted and changed the economies and social structures drastically for the better in those societies which have massively applied computers. This comes about because computer technology and computer applications are self-nourishing. It creates new technology and resources through its application, which are needed for its continued expansion and the growth of the area for which it is employed.

When examining such trends, futurists are struck with the similarities of emerging computer systems with biological systems. These partial analogies are simple; perhaps this is a major reason for their appeal. The analogies do not give us simulatable models of large computer systems because we do not know how to simulate such complex analogous biological systems. Nevertheless, such analogies are helpful, since we do know a great deal about their functioning.

Since logic components are becoming smaller, faster, and more complex with the advent of LSI (the large-scale integration of electronic components), we find we need to move control deeper into the guts of the system. This means that we are forced to transfer codes from the centralized control system, which are to be remotely decoded, to the distributed control points located locally with the remote functional groups of logic, where the desired operation is to be performed. Such transfers of control are not too unlike the RNA transfers of information from DNA to remote points. As higher degrees of circuit integration occur, system designers are predicting the need for more and more code transfers.

In some designs of future computer architecture, internal computing engines are beginning to merge. In these systems, we see an extension of such RNA-like behavior in which logic codes, as well as total programs, are being transferred and decoded at remote but internal computing points. Designers find that fewer control lines (channels) are required by transferring codes to be decoded locally than when the logic codes are decoded centrally, for this requires that many tentacles of control emanate to the local functions. Local functions requiring local control include such internal functions as arithmetics and memory, but are far removed from the centralized control functions (they are not on the same LSI waffer). As it turns out in actual LSI computer design, the LSI waffer's most severe limitation is its limited number of connection pins. By

doing RNA-like transfers, the number of pins per LSI waffer
tends to be minimized. Obviously, by reducing the number
of pins required, the number of wires that connect to the pins
are also reduced. These wires contribute substantially to the
bulk of such systems. By reducing their numbers the total
length of interconnecting wiring is reduced and the system is
capable of faster operation. Further, one of the most unreliable
parts of a large computer system is the individual connection;
by simply reducing the number of connections required, the
system can be made more reliable. There are many compelling
reasons to draw analogies between the RNA transfer process
we encounter in biological systems and next-generation computer systems in order to arrive at innovative new solutions
to today's design problems.

In very large systems that involve complexes of computers,
we find a growing tendency toward another form of transfer
which again smacks of RNA transfer. It turns out that the
documentation of large programs is unable to keep up with the
changes to such programs if a "hard copy" (printed) of the
program is required. The tendency is to let the computer
system display the current status of the program in its memory.
In the past, when a number of computers were to use the
same program, it was *hard copied* on cards or magnetic tape
and sent to the remotely located multi-computer system. Today,
we are finding this process too slow and unreliable and are
beginning, in some systems, to transmit the programs electronically from a central system to the other remote systems.
This is done every time the program changes.

A technological extension of this process is being investi-
gated and comes under the title of "paperless books." What
is meant by paperless books is the total production, dissemination, and display of magazines, journals, and books by electronic methods. The process starts with the initial draft of the
manuscript by the author either dictating, writing or typing
his/her thoughts directly into a type of remotely located TV-
like computer terminal. The computer then edits the material
—acting like a combined secretary and technical editor it

corrects spelling, suggests better sentence structuring, calls out repetitious material, suggests reorganization of thoughts, abstracts, collects key words, compiles citations, and builds descriptors. The system plays it back to the author, in real-time, and together they produce further drafts. The resulting tapes (manuscript, cataloging information, bibliographical information, and abstract) are sent directly to the publisher when the author is satisfied that he has a final draft. Publisher is used loosely here, since no hard copy production is anticipated. The publisher is the distributor of the information to libraries and those professionals who desire it and also the one who determines the worth of the so-called publication. Most desirably, this information would be sent via wire to other computers located at major dissemination points (large cities, for example) from which users have on-line access to the information. In fact, these major dissemination points would be set up in a current awareness fashion. That is, upon receipt of new information, they would search their profile files of users to determine if they should make a user aware of any specific item that is a high priority item to him. Most users, however, would make use of these centers on a request-query basis by searching the files as they need information via a remote terminal.

The user, upon being led to a particular item through a tree search at a terminal with display facilities, can then read the abstract to determine if he is interested. If he is still interested, he can call up textual material and scan it for the information he needs immediately. If he desires a hard copy, he can request a printed copy from the system. However, few would, because most would satisfy either their need for information or their curiosity at the terminal. Thus, the chain of events that occurs from the origin of an idea to the eventual user is considerably shorter; it can be handled electronically from the moment the author conceives his ideas! Further, the cost of getting the information to the user, because of the reduced steps, could be greatly reduced.

Note: Throughout the described process, no hard copy (on paper or otherwise) is produced (unless requested); hence the name "paperless" books. Can you imagine the savings to all who produce lengthy subject bibliographies, and to the many technical libraries who produce abstracts and catalogs and retrieve information? Just imagine the amount of duplication of effort that can be eliminated. What about the savings in file storage and book shelf storage? Isn't it possible that these savings can more than pay for implementing such a system in the future without even considering the increased ease and speed with which information can be disseminated?

Obviously, we need to wait awhile for such a system to be in widespread use. Today CRT (TV) terminals and mass storage of data are both too costly, but someday, in the not too distant future, the economic scales may be tipped in the favor of such a system.

With the advent of time-sharing systems and LSI, we can determine a trend toward "Utility Conglomerate" computer systems. These utilities will supply information to many on-line users in much the same fashion that power is supplied, but with some major differences. As logic becomes lower cost, it is finding more and more use in the remote terminals of such information utilities. Today these terminals are too costly, but fortunately optimistic projections of their ultimate future low cost are predicted. As they are adapted to more and more applications, the desirability of having more capability in the terminal increases. LSI introduces the era of the so-called *intelligent* terminal. Designers are predicting that future terminals will perform such intelligent operations as:

— Current awareness
— Filtering
— Sorting
— Disseminating.

The current awareness function will filter out extraneous or duplicated information at the extremities in the remote terminals much like the function is performed in the eyes and ears in the biological system. As new or desired information is encountered, it will signal the CPU (central processing unit) by an interrupt and then transmit the information, i.e., make the CPU currently aware of the information.

In accomplishing this current awareness function, some of the computer's capability is moved to the remote terminals—moved to the data source and collection points as well as to the data/information dissemination points.

The current awareness function also works in the reverse fashion for the computer systems that are connected to our communication systems. Perhaps the most used terminal of a future computing system complex will be the telephone. When the computer obtains or processes the information that a user needs immediately, it will be capable of making such a user currently aware by telephoning him. We have had this type of capability (such as voice drums) in some military systems for about fifteen years; it is only a matter of diverting this technology for the use of society in general. In such a computer call-up system, the computer would piece together the message from an inventory of words, sentences, and phrases contained in its memory. Research to find ways and means of enabling computers to interpret voice patterns is continuing at a rapid pace. This research will result in the eventual communications link for use in the opposite direction. In the meantime, we can use the touch tone type telephones which the Bell System is rushing to convert to in the early 1970's.

In those multiple-computer command and control systems that are beginning to emerge for both military and commercial applications, we begin to see hierarchial structuring. This layered structuring is somewhat like structured societies (for example: local government, state government, and federal government, or supervisors, managers, and directors). These systems are controlled from the top level by the central (headquarters) computer. Direction and supervision is fanned out from the central computer to the local computer centers that, in turn, provide the detailed and specialized control needed by the local remote terminals and computers.

Still another greatly over-simplified analogy which compares the computer to the biological system of the brain can be found in some multi-processor systems. In such systems, separate processors for mediating different functions are coming into common use. Some of the processes that are mediated (processed or computed) in parallel with other functions in multi-processor systems are:

— Executive control—the system controller; its operation and function is sometimes compared to the hippocampus
— Compiler operation and emulation—language translation
— Application cooperation
— Data processing and data reduction
— Input/output—communications.

The major reason for offering these analogies is to gain a better understanding of the lesser known computing systems by borrowing from our practical knowledge of the biological systems. In many cases, by simply making the analogy without further analysis, the designer is led to new innovative approaches to computer system design and architecture.

Problems in the Management of Large Computer Systems

The program stored in the computer's memory, which is transferred to the CPU for execution, is the control mechanism by which the computer makes decisions before performing the desired processing of data and computations. Without a program, today's digital computers perform no practical operations. The program which directs and manages the operation of both the hardware and other programs is the Executive program. Today, this program encounters difficulty in the management of large systems; in other words, the designers of Executive programs need to learn more about how to manage systems. A partial solution to the problem seems to lie in the marriage of hardware and software by building-in some special hardware to assist the executive and some hardware for performing tasks of the executive which heretofore have been performed only by the software.

Perhaps the biggest problem facing the user of computers

and the computer manufacturer alike is programming. When researchers look at how fast we are hiring programmers, the statistics show that it is one of the fastest growing industries. In fact, we are hiring programmers faster than the population is growing. The following story draws an analogy with the telephone industry. In the embryonic days of the telephone industry, a researcher discovered that they were hiring telephone operators faster than the population was growing. The communications scientist's conclusion was that everyone would have to become a telephone operator! If we think about it, we realize that everyone who uses a telephone has in fact become a telephone operator, with the advent of the dial system. Of course, to achieve this advance, the telephone industry added a lot of electronic gear in its central offices. The computer industry must do much the same thing. Today it is doing the R & D necessary to design the electronics needed in the CPU to allow the user to "talk" to the computer in the user's language so that he can program the computer directly.

Of the total amount of money being spent for all data processing, the percentage being spent for software has changed drastically from the days of the first computer. It has increased from about five per cent in the early 1950's to about seventy-five per cent today. Such a phenomenal growth cannot be expected to occur without problems. The problem cited most frequently by managers of programming is that they lack methods for predicting and controlling programming costs and schedules.

In the use of a computer, the neophyte tends to believe that the program which performs the operations of the application for which the computer was intended should be larger than the programs which support and control the operation of the system. This is not the case today—the so-called system programs, in contrast to the application programs, account for more than fifty per cent of the computer's storage and programming effort.

The Executive control program; the program which automates, manages, and controls the operation of a total computer system, including all the resources of the hardware and software of the system, is one of the largest system programs. In the past, when the computer system failed, the programmer blamed the hardware and the computer maintenance man blamed the program—and a battle royal often was the result. Today this is not the case. The computer has become quite reliable and seldom fails. Now when a computer fails, it is the operation system, usually the Executive, that is blamed. Some of the problems with today's Executive programs are:

1. The complexity and the size of Executives are growing at a rate of about ten times each five years.
2. The requirements for main storage to hold the Executive has grown until it exceeds reasonable bounds. In 1964, a typical large-scale system's Executive required 10,000 words of resident storage; today it requires over 50,000 words.
3. The total operating system (the collection of system support programs including the Executive) is now greater than a million words for a typical large system.
4. To the manager of the Executive programming group and the user alike, the Executive is a living, unstable, expanding, and contracting organism in the way that it constantly and dynamically changes.
5. The overhead in computing capability required to execute the Executive, due to the many new demands made for its service, has grown until it is out of control. Users want to have many general-purpose tasks performed by the Executive and new requests are coming in faster than these additional tasks can be efficiently incorporated into the system. The result is that Executives have become too large and complex to write and debug.
6. It is difficult to recover when the Executive "bombs out"— and Executives fail all too often.
7. There is no standard language for writing Executive control programs.
8. There are no standard measures of computer system efficiency. The general feeling is that we need to get 100 per cent efficiency from computers, but no one has yet defined what is meant by efficiency. For example, it is relatively

easy to keep the CPU busy all of the time, and to some this means that the system is 100 per cent efficient. Most others realize that if much of the rest of the system is idle, even though the CPU runs 100 per cent of the time, the system is not effective, since the CPU is probably the smallest single portion of the system.

In the general area of programming, some major problems are:

1. Computer applications are growing faster than the supply of available programmers and programmers cannot be trained fast enough.

2. Industry has been forced to use untrained programmers which compounds the management problem by as much as a factor of ten, i.e., it requires as much as ten times more programming time to program a large-scale task using inexperienced programmers.

3. Almost everyone believes that we are further along than we really are; programs are not yet available for every application.

4. The debugging of a program and integrating it into the system may consume as much as seventy-five per cent of the total programming time for the task.

5. Programs change so fast that program documentation fails to keep pace.

Trends Toward the Future

A major event that many experts predict will occur during the 1970 decade is that computers will, at the least, take over all of the tasks that most people believe they perform today.

A typical large computer system is made up of four major hardware sections when divided according to its functions:

— CPU—the Central Processing Unit usually called the computer but not the computer system

— Memory—the memory consists of many hierarchies

— Peripherals—the peripherals comprise the auxiliary memory and input/output communication devices
— Terminals—the terminals are the man/machine communication interface.

A typical current scientific *number crunching* system size comparison of these major functions in terms of cost or number of components (logic gates and bits of memory) is:

— CPU, 25 per cent
— Memory, 25 per cent
— Peripherals, 50 per cent
— Terminals, Nil.

A similar hardware division comparison for a typical future information *data manipulative* system is:

— CPU, 5 per cent
— Memory, 10 per cent
— Peripherals, 20 per cent
— Terminals, 65 per cent.

In either the scientific or information system, however, the software accounts for more than fifty per cent of the cost.

The cost trend in the logic [4] or the hardware of computers is:

— Before 1960—about 2 dollars per logic switching gate
— Today—10 cents or less per gate
— In the early 1970's—less than 1 cent per gate.

Since 1960, the cost of logic has decreased sharply and its cost is expected to drop to one-tenth its present cost within five years.

Since there are problems in programming, researchers of next-generation systems are today investigating high cost problem software areas with the idea of replacing some of this expensive software with inexpensive hardware. This will, of course, increase the cost of the hardware. Since hardware logic represents considerably less than fifteen per cent of the total system cost, a slight increase in the cost of hardware will not materially increase the total cost of the system. In fact, if the R & D which is now going on is successful, the total cost of the system should decrease by a considerable amount. The types of program functions that are being studied for inclusion in the logic to solve today's problems in future systems are:

— Primitives—commonly used subroutines which are required for the programming of a variety of applications
— High level language—languages which are more adaptable to man
— Memory management aids
— Features to ease the executive control and system management problems
— Debugging aids

Today, the program designer and future computer system designer is instrumenting programs, much like using an oscilloscope on hardware while the programs are run dynamically. The program instrumentation process gathers statistical data on the use of such system resources as subroutines, registers, memory, peripherals, and time. Recently, a discovery was made at Univac when data from a number of instrumented runs of a compiler program were being analyzed. The analysis of the data gathered showed that there are four basic subroutines which represent approximately ninety per cent of the operations performed and account for fifty per cent of the run time of the compiler. Three of the four subroutines were list (table) operations. These are:

— PUT—operand to top of list
— TAKE—operand from the bottom of list
— MOVE—operand to bottom of list.

The fourth subroutine was a one instruction type—branch unconditionally. It should be noted that in any compiler system, there is also a rather lengthy calling sequence to obtain a subroutine. From this analysis, the program designer obtains the information needed to optimize the compiler, increase the speed of its operation, and make it more efficient. The system designer obtains the information needed to build next generation systems using hardware for parts of the program in order to ease the programming problem and make the system more efficient. Design studies indicate that by building in such features, which often will require little additional hardware, operation of such a computer system will be greatly enhanced. It turns out that these same list operations speed the operation of such other programs as Executive control, information retrieval, management information systems, and file maintenance by an even greater amount than they enhance compilers. Such list operations fall into the class of primitives.

Tomorrow's solutions for the Executive control program will reduce its size and overhead by more than fifty per cent and yet make it twice as capable.

The computer memory is also undergoing radical changes. In the 1950's the cost per bit of the computer memory (random access and non-rotating) was a dollar or more. In the 1960's the average cost per bit has been reduced to about ten cents per bit for large systems and by the mid 1970's the cost per bit should be reduced to one tenth of a cent per bit or less. The cost of main memory, which is usually magnetic, has dropped until its cost per bit is less than one tenth the cost of a comparable memory ten years earlier. In the 1970's, the cost reduction rate of memories should approach the cost reduction rate of the logic. Today, much of the magnetic memory consists of logic and thus the cost and capability of a memory system is paced by the logic. Further, in the 1970's, many memories will be made up entirely of logic (semi-conductors). Therefore, it is not too surprising to find that memories will follow the same cost trends as logic.

The speed of the on-line random access memory has similarly undergone a major change: its speed has increased. The following tabulation shows the trend of typical large computer system memory speeds.

— Mid 1950's—10 microseconds
— Mid 1960's—1.0 microsecond or less
— Mid 1970's—0.1 to 0.001 microsecond

Note the expected speed breakthrough in the 1970 decade, instead of a speed improvement of a factor of ten each ten years, we can expect a factor of ten increase each five years.

Not all of the increases in memory or computer system speeds are gained simply by employing faster components. The system designer is beginning to apply his technology through architectural changes. For example, large high-speed memories are costly. To bypass this problem, computers are using a hierarchial memory structure which employs a "speed or gear changing" memory. In such an architecture, a relatively small, high cost, very high-speed memory is coupled with a low cost, relatively slow, very large memory. In this arrangement, the design allows for block transfers to the high-speed memory in a fashion that for a given time thereafter enables the CPU to find all of the next references to memory in the fast memory. The result is to effectively make the total memory system appear to the program as if it were high speed. This and other similar architectural breakthroughs will allow memories of the coming decade to appear to the user as if they are much faster and will account for much of the predicted speed increases of future memories.

The typical capacity of an on-line random access memory for a large computing system has also changed radically through the years and the increase is expected to continue for the coming decade. The trend in memory capacity for large computer systems is:

— Mid 1950's—100 thousand to 1 million bits
— Mid 1960's—1 million to 10 million bits
— Mid 1970's—10 million to 1 billion bits.

Today, the total memory capacity of a single computer system, including its on-line auxiliary serial access storage, is limited to between 10^{12} and 10^{13} bits, with a ten- to a hundred-fold increase expected by the mid 1970's.

Large associative memories are looming on the horizon—memories addressed by content. Today, these memories are small and expensive. They cost about fifty cents a bit for a 100,000 bit memory. By the 1973 to 1975 period, associative memories will grow in size to one hundred million bits and will become economically practical at prices of one to ten cents a bit. Such associative memories will allow Hebbian type nets (dynamically reverberating) to mediate, like the brain, performing many operations in parallel and enhancing information retrieval.*

The growth of computer power is increasing by a factor of ten each five years when measured in terms of the instructions executed for each dollar spent. In 1960 we were able to execute about one million instructions for a dollar; by 1970 we can expect to be able to execute the same number of instructions for about a penny. If we look at computer power in terms of the number of bits processed, a factor of ten increase occurred in the past two years.

With the advent of the computer terminal, with many terminals coupled remotely to a large time-shared computer system following the utility concept, the cost for using a computer to perform a simple useful task can be measured in pennies. Such systems are making the computer's power economically available to many people. In the 1970 decade, it is expected that the computer's power through the utility concept will be made readily available and its use will become widespread, like the telephone.

The programmer or computer engineer, buried in his work, becomes solely concerned with the facts and figures of his tasks. He is motivated by a desire to record such information accurately to produce products and not by the impact of his work

---

*Editor's note: In this connection the reader might be interested in the cell assembly theory of mind to be found in: D.O. Hebb, *The Organization of Behavior* (Wiley, 1949). There are also some interesting comments regarding the idea of reverberating or circulating memories to be found in: N. Wiener, *Cybernetics,* 2nd edition (MIT Press, 1961) pp. 146-147.

and technology upon life in society. Some computer systems designers and managers, unlike others who preceded them, are beginning to be concerned about the computer's impact. As a result, they are specifying safeguards against both the unwitting and intentional wrong use of next-generation computer systems—they are adding features to encourage computer usage for the improvement of society. Today, since little has been done in this area, the designer is limited only by his own imagination. By using this license intelligently, the designer is creating meaningful new technological adjuncts which enhance the spread of computers for applications in more ethical and social areas.

The number of computers per million people in the United States is more than 300 and is increasing; there are about 50,000 computers in use commercially today (not counting those in use by the military). In most foreign countries, there are fewer than 100 computers per million people. Is there any wonder that the technology gap exists and is widening? With the advent of the remote computer terminal and the immediate response systems becoming a reality, the availability of the computer is undergoing a drastic change. The computer and its power is available to anyone who has access to a remote data and information communication terminal. When we look at the statistics from this viewpoint, today's computer availability per million people in this country is estimated to have more than doubled. By the mid 1970's, the projected computer availability for the United States should approach a terminal for every 100 persons.

The rapid proliferation of computer terminals, if one includes the telephone as a computer terminal, heralds a tremendous impact that is just beginning to be realized. The use of remote terminals has just begun.

The number of people required for the operation of each computer decreased by more than a factor of two in the last five years, while the number of computers in use has quadrupled. Any reduction of personnel required is a myth, since actually twice the total number of computer personnel was required. It is interesting to note that the reduction of people was primarily due to the reduction in the number of key punch operators required, which dropped by a factor of ten.

This trend has nearly bottomed out and no further reduction of any particular consequence should be expected. This leads to a problem: where are all the people who will be needed to work on computers coming from? One answer is to find ways of reducing the number of programmers required. Research toward this end is being conducted in both hardware and software.

One approach to making computers capable of being programmed by the user is to build in higher level language capabilities. Since language changes dynamically and is specific to each application, it will be necessary for future computers to be both capable of being: (1) conditioned to a desired language, and (2) adapted to the special language requirements (semantics) of each user.

This is the direction that computer designs for future systems are taking—toward language-enhanceable processor systems.

This is not too surprising if one draws a correlation between the development of computer language and the development of spoken or written languages which society has used for centuries. All languages (including computer languages) develop along the same lines. They begin with a limited number of different words—early day computers had fewer than 100 simple instructions. The second stage in language development occurs when considerable complexity is added to the few words in the language. Latin is an example of such a language at this stage. Most modern day computers are also at this stage of development; they have approximately 100 instructions with as many as 30,000 variations. If a language stops at this level of development, it eventually will become a dead language for most practical communication purposes. Again, Latin is an example. To determine the third and final known stage in language development, one need only inspect a modern dictionary. It occurs by getting rid of some of the complexity but not all of it, and the almost unlimited dynamic addition of new words as required. This stage of computer language development has begun in the software areas through compilers and is being researched from the hardware viewpoint at the present time.

The problem of programming, if this last language development stage is not undertaken, can be shown by the following analogy. Suppose that before this paper was begun, the restriction had been made that no more than 100 different words could be used. The size of this paper would have expanded until it would be larger than an average book. If it were completed, it would be far too complex to be understood without a great deal of study. It would require an awfully long time to write such a volume. In fact, it probably would never have been completed.

The management of large systems, such as companies, government, and society, is no small problem today. There is great hope that computers will contribute to the elimination of future management problems. To this end, there are many Management Information Systems (MIS) being set up to gather and sort the information needed for managers to make decisions. The trend in these MIS systems is toward Management Information and Control Systems (MICS). Such systems have at least the properties of:

- Real-time
- Time-shared utility type system
- On-line multiple remote connection to managers and users through the computer terminal or through other sensors and effectors, especially the telephone
- Current awareness operations—the system makes the user aware of items of specific interest, rather than either flooding him with all information or requiring him to query the system
- Integration of total operations, including administration, planning, management, technical, design, manufacture, sales, and customer aids
- Information retrieval—the availability of low cost mass memories is making information retrieval economically feasible

— Data bank and information distribution system
— Profile files on each individual's interests and needs which are dynamically updated with use.

An example of how such a system might control its environment can be seen in the mind's inscape by considering the following situation. Suppose I program a computer to process this week's PERT (a management planning tool: Program Evaluation and Review Technique) information. The computer reduces the PERT information and determines that my group will be thirteen weeks negative (i.e., will be thirteen weeks late) unless I solve my scheduling problems. I again send the computer PERT inputs the following week; if I try to cheat by not calling up the computer, it will run the PERT information and assume there is no change! Suppose in the last week, I make little improvement and the computer determines that I will still be late by an unacceptable amount. This time the computer will call both me and my boss! Thus, one of the mechanisms available to the computer through the MICS program for controlling its environment becomes visible. Next week, if I still haven't solved my problems, the computer will call me, my boss, and his boss.

Some will look at this in the Orwellian fashion and feel that "big brother is watching you." Others will see that the opposite is true; it will alert managers in real-time early enough so that they can help. In the above example, it would allow my bosses to help me by doing the managing thing which prevents me from being fired and helps get me back on schedule or perhaps changes an unrealistic schedule. Those of us working in the MICS field are predicting that such systems certainly will evolve and come into mass use. Further, we predict that those managers, companies, politicians, and governments who do not use such systems by the 1975 to 1980 period will simply not be able to compete in a society which does.

In addition, future computers will contain many other new features, commonly called bells and whistles. And as always, some will prove to be useless and others will lead to new problems as time marches on.

Conclusion

No fact of contemporary life is more challenging to society than the present rapid advancing technology and continual cultural change. These changes are revolutionizing the computer and communications industries and have set in motion irreversible processes which affect the whole structure of society. We could never go back to the pre-computer era; our dependence on computers is now far too great. Not only are the material and economic conditions of life enhanced and affected by this advanced tool of the cybernetic age; there are also far-reaching changes in our moral values, our politics, the way we learn and educate ourselves, and our art and intellectual thinking.

If this reach for the future seems a radical departure, it is because the potential of next-generation computer systems, in a cybernated era, is so great. The demands on what we do now dictate that we take radical measures to bring about the future we want *now*—not later.

The headlong rush of progress in computers amidst the cybernetics revolution has been somewhat overshadowed by an advancing communications technology which has partially hidden from view what computers have been doing for mankind and his problems. Vast strides have been made in communications through media augmentation and new technology, despite the fact that we are still faced with the problem of communicating information instead of data. The problems of real communication are growing worse day by day, especially in handling dialogues at all levels. We also have problems with computers. Can it be said that we are really at the threshold of the cybernetics age, an age when we solve our problems and use our technology to control and communicate for the benefit of society? Perhaps not, but we can at least say that we are producing some of the tools, such as the computer, which will be necessary in such an era.

This paper grew out of the research I performed in order to advise management on the needs for future computers and my personal concern about the computer's impact on society.

Its object has been threefold: (1) to pose some questions not frequently asked by computer system designers about our present and future situation; (2) to indicate the very real problems in the management of large computer systems and show how critical the need to solve these problems has become; and (3) to suggest the solutions that may be instituted to alter the physical properties of future systems. In addition, I tried to touch on the computer's impact on our future social environment.

I have tried to put on tomorrow's glasses in order to see where we can go, rather than wear yesterday's glasses and only be able to see how bad it was. Because of the high cost of programming and the need to capitalize on its investment, new advances and ideas will tend in the future to rest on the shelf much longer than such advances have in the past. The state of the economy that surrounds programming will govern our actions more than we would like; it will inhibit our ability to advance into the future as promptly as we know we could. Thus, there should, of course, be some skepticism concerning the certainty of these predictions. The era that we thought we had entered when yesterday's issues first sparked heated controversy, and then abruptly dropped in importance as even more revolutionary changes were thrust upon us, now appears to have been altered into an era where today's problems may be with us for a long time to come.

REFERENCES

1. Earl C. Joseph, "Computers Trends Toward the Future." *Proceedings of the IFIP Congress '68, Edinburgh, Scotland, 1968,* Invited Papers Section, pp. 145-157.
2. Hoyt Ammidon, "Communications Revolution: But Is Anybody Listening?" *Vital Speeches of the Day, 34,* 1968, pp. 469-472.
3. *Daedalus, Journal of the American Academy of Arts and Science,* Special Issue: "Towards The Year 2000: Work In Progress," Summer 1967.
4. E. Bloch and R. A. Henle, "Advances In Circuit Technology And Their Impact On Computing Systems." *Proceedings of the IFIP Congress '68, Edinburgh, Scotland, 1968,* Invited Paper Section, pp. 24-30.

Cybernetic Analyses of Large-Scale Computers

C. J. PURCELL

Control Data Corporation
Minneapolis, Minnesota

I have to thank Dr. Kemeny for introducing this group to my subject: large-scale computers. I trust there is room for a cybernetic technician amidst all of this cybernetic philosophy. I find that there are fewer philosophers present than I had previously imagined.

The place of the technician is to analyze and synthesize in order to provide useful relationships. Normally I would call this useful relationship a "system"; however, the word is so overused that I now try to avoid it. Many other words heavily employed by computer technicians must be defined before use.

Cybernetics, at best, consists of the development of useful biological, social and physical relationships. There is much to be done in this whole area. Cybernetics should allow us to build on the shoulders of our predecessors—not destroy them.

At this instant, however, we will examine a physical relationship of interest called the large-scale computer. This computer, if properly organized in the context of utility, may prove to have some influence on social and biological relationships as well. The design of large-scale computers still resembles an artistic effort, not an analytical effort.

This large-scale computer is to provide interrelated computational facilities not now available. In order to insure and verify that the large-scale computer can be useful, I took the time to review once again the key book by Norbert Weiner, *Cybernetics*.[1] This book has an amazing overview of the place of the digital computer in a cybernetic society.

The most comprehensive cybernetic machine indicated by Mr. Weiner has not yet been made available, but I believe that this machine, the communal information center, can be in heavy use by the mid-seventies. We shall develop the specifications of such a machine by reviewing digital devices for information processing in order of appearance in Mr. Weiner's book.

Communication considerations occupy the primary position in any digital system. We find that noisy data must be made smooth. Readings of sense organs, human frailty and circuit malfunction all contribute to noisy data, while on the other hand the digital computer is absolutely unforgiving of error. The output of the digital computer must be useful to humans, at least, if not in a format suitable to control another machine. In a simple sense, redundancy is the cybernetic solution to noisy data. Simple redundancy is required in outputs. This redundancy adds to the size and workload of the large-scale computer.

Cybernetics postulates a variety of learning machines. I have seen very few successes in this area whereby the digital computer is forced to mimic a given human-like capability—people are much cheaper. On the other hand, the digital computer has been most effective at adaptive processes operating at time scales much slower, or faster, than the human can react. An adaptive missile tracking system can produce fantastic performance over any manual tracking system. Another type of learning is illustrated by the slow accretion of useful procedures and algorithms in the digital computing community. These procedures and algorithms enable the computer organism to adapt over a very long time span and provide a constantly improving service.

Cybernetics has much to say about computers. Mr. Weiner obviously prided himself on the following set of predictions, made in (approximately) 1942, which concerned the nature of computing and which still hold. The computer should be:

1. Digital and not analog.
2. Electronic and not mechanical.
3. Binary and not decimal.
4. Automatic and not manual.
5. Alterable memory and not fixed memory.
6. Boolean in implementation.

All of these six properties are now in use in successful digital computers. Very little justification can be provided for changing from these suggestions, no matter how hard we try. A current thought on statement 4 would also allow a substantial amount of interreaction with the computer during the computing process. When this occurs I feel that possibly the mathematical theory of the specific situation under study is poorly understood.

In practice, we find that a typical large-scale computing task is built up of three equal parts. These include:

— ⅓ of time is spent in number crunching (the DIGITAL MACHINE)
— ⅓ of time is spent in deciding what to do next (the LOGICAL MACHINE)
— ⅓ of time is spent in processing input data and preparing output (the TEXTUAL MACHINE)

Each third of the computer must be very capable in order to accomplish the assigned tasks.

The digital machine contains provisions to perform a variety of simple arithmetic processes. In addition, a variety of useful, but complicated, built-in techniques to aid the technician can be provided at little additional cost, which will enable structured data to be processed very rapidly. The logical machine is useful both in the sense of problem solving (a decision tree or existence table) or in the sense of aiding problem solution (instruction branches).

The textual machine must accept large streams of input data which in turn must be dissected into meaningful source tables. After processing, the textual machine must efficiently form useful output in graphical, or tabular, form for visual consumption. Acoustic, pictorial and control information also require test manipulation during problem solution.

I have described a collection of willing slaves now employed in the service of mankind. The general application of these slaves can be seen in many existing computer centers and computer control applications. An aggregate of these slaves was also described by Mr. Weiner. This aggregate is defined as the communal information center. The function of the communal information center is described by its capability of modifying the behavior of a specific individual in response to the actions of another specific individual. This function proves its requirement for existence, but also points to its greatest peril.

The general properties of a communal information center include:

— Perfect memory of infinite size.
— Boolean logic.
— Binary implementation.
— Noisy inputs.
— Compact outputs.
— Public, privileged and private sharing of information and procedures.
— Workspace for each user.

This center is a new organism within society. Such centers will be located at geographically convenient points, as well as interconnected on a regional or national basis. The communal information center is not just a remote calculator located at the end of a long string tied to your finger, it allows for organized and specific communication between individuals through the use of shared procedures and shared data in public, privileged or private entries. This interreaction between individuals without personal contact will provide great power to the community as a whole. The communal information center as an organism will live on forever by adding, modifying or purging its information content. I quote Mr. Weiner: "Any organism is held together . . . by the possession of means for acquisition, use, retention and transmission of information." [1]

Our goal is to implement this communal information center in the early 1970's. While the cybernetic abstract can be very exciting, the cybernetic reality must be useful. It appears that the following functional and performance requirements on the communal information center can be fulfilled in the early 1970's, in order to synthesize a useful facility:

1. Most important, compliance with the graphic and information interchange standards of the American Standards Committee in eight bit "byte" format (especially here at the National Bureau of Standards)
2. Provision for connection by dial-up procedures of some 10,000 telephone digital links simultaneously, via a private communications' exchange, or public "CENTREX" telephone exchange
3. A trillion bits of random access interconnected memory, capable of supporting the 10,000 users via directed entry in public, privileged or private memory, as well as a large supply of personal work space, to make possible the fulfilling of a small, but useful, request in no more than ten seconds.
4. Entry to procedures which consist of simple library functions, as well as disciplined procedures which can be attached together to perform any unusual task, can be public, i.e., open to all, or private and open only to a few individuals for a specific function. Languages can be an entry into procedures and not, we trust, an effective bar to keep out all but the professional.
5. Central logic which will enable efficient processing of requests in the context of the many disciplines which utilize computers—or as organized structures. These include:

— Ordinary numbers within range of man's awareness
— Fractional numbers outside of normal range
— Very large numbers of many digits for curiosity if nothing else
— Non-numbers.

Examples of non-numbers include all sixteen Boolean operators. Left aligned alphanumeric character strings must be processed for text manipulation. Data must be obtained from files; however, data that is in place should not require explicit *rules*. Thus, an implicit input/output scheme will simplify the organization of useful programs. As time has passed, our users have become much more knowledgeable of the most applicable standard digital techniques to solve a wide variety of tasks. The central logic must provide this representative set of standard functions, while not inhibiting creativity toward the development of new techniques and procedures.

6. The reality of cybernetics demands that a cost of the communal information center be determined. While this cost is difficult to state, one can set an outer limit of such a system suitable to furnishing central service on a continuous basis. This cost should not exceed $20 million.

7. The costs of operation and training must also be considered. In order for our organism to stay healthy, we must care for the physical components, as well as maintain and update the information components. In addition, the training procedures to develop facility in the use of the information center may be easy to implement. The massive set of on-line terminals will all be able to provide a tabular answer to any procedural question, as well as provide for the classical case of computer-aided instruction.

The resultant communal information center will possibly provide news, statistics, periodic voting methods, a checkless society, adult education, recipes, information responses to inquiries, status responses, directives, large-scale computations, police facilities, community planning support, security monitors, public health monitors, gaming resources, libraries, and, finally, dating bureaus and employment agencies.

We can picture the center as consisting of many tentacles connected into business places, homes, warehouses, etc. and interconnecting these locations in an adaptive centralized frame, as suitable. The regional information centers will provide the new global entries into the commune, while collecting and responding to specialized information and processing requirements.

This vision of the future is attainable using today's technology with small extensions. On balance, the potential good will far outweigh the potential evil. If enough people, say one million, all share access to a communal information center, the cost may be only one dollar per month each or less.

REFERENCE

1. Norbert Weiner, *Cybernetics*. (Cambridge: The MIT Press, 1948, p. 161). Reprinted by permission of the MIT Press. Copyright by Massachusetts Institute of Technology.

Large-Scale Information Systems

ARTHUR G. ANDERSON and MICHAEL E. SENKO

IBM Corporation

Introduction

One of the hallmarks of the scientific era is man's capability to mobilize great numbers of personnel to work swiftly and efficiently with a single common purpose. The purpose may be the landing of a man on the moon, more efficient utilization of the earth's resources, or the optimum fulfillment of some consumer demand. Whatever the purpose, existing cooperative groups have grown to such size that traditional methods of obtaining, correlating and transmitting information are no longer adequate. Large-scale computerized information systems are essential in the present environment and they appear to provide the only means for creating even greater cooperative efforts.

Our talk shall look, to some extent, at the technological details of computerized systems, but we will be more concerned with the effects that the implementation of these systems has on the individual, the group and the total project.

109

Technology

The first individual to be discussed is the constructor of the information system. To understand the changes in his environment, we must investigate the rapid evolution of computer hardware and software technology.

Hardware

Few industries can match the computer industry with regard to improvement of product cost-performance. From machinery like that installed in the Columbia Statistical Bureau in 1935 (Fig. 1) to present day equipment, the cost per primitive operation has dropped by orders of magnitude. This

Figure 1.

progress, however, has much less effect on large-scale information systems than it has on scientific and engineering calculations. Information systems are more dependent on the ability of the computer to store, provide access and transmit billions of characters of data in a real-time mode. Fortunately, major progress is also being made in these functions. At the present time, we can get access to any one of 100 million characters within a tenth of a second. By the 1970's, we should be able to obtain similar or faster access to billions of characters on a single device. In fact, trillion bit memories with access times of five seconds have been delivered for certain forefront applications. When one considers that this last file has capacity equivalent to a stack of punched cards 2,500 miles high or to twenty acres of file cabinets, he begins to understand why unaided humans cannot cope with the volumes and high activity rates of large-scale information systems. Large-scale information systems are indeed a quantum jump in man's ability to deal with information.

Similar progress has been made in the area of communication with computers. In 1957, the first work began on the use of terminals in a large-scale commercial information system. Since that time, thousands of terminals have been installed to provide rapid communication with large data bases. Companies have been able to take advantage of the decreased costs associated with expanded use of communication lines (Fig. 2). Accompanying the spread of communications utilization in geographically distributed systems is a change in the role of terminals. Figure 3 lists the trends which are occurring. Of course, we also see more use of cathode ray tube terminals, potential in the use of non-impact printing terminals and a substantially greater trend to interactive graphic modes of operations.

It is, of course, up to the systems constructor to pass on the benefit of these achievements to his user. His most pressing problems are in the software area.

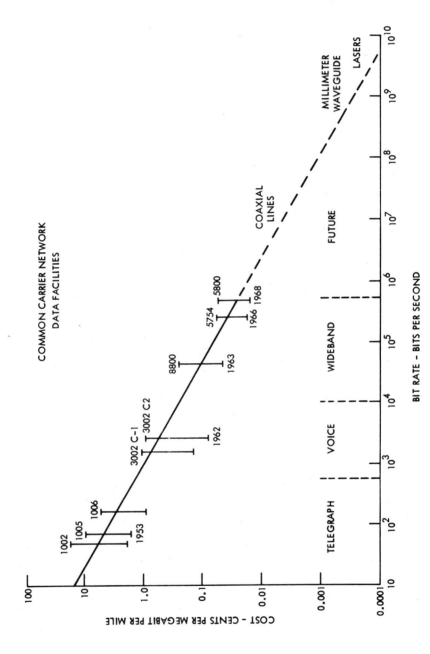

Figure 2.   Common carrier network data facilities.

- Lower costs per bit
- Higher speed facilities
- Wider variety of transmission facilities
- Significant geographic expansion of all services
- Wider variety of interface hardware
- Significant shift toward digital transmission and switching
- Standardization *and* flexibility

Figure 3.   Communications trends.

Software

There are both qualitative and quantitative trends in the software area. For purposes of discussion, there are two sub-fields: systems programs which the systems constructor provides as a base for all applications of the hardware, and applications programs which perform a specific user-oriented task.

In this connection, there is some analogy between scientific study and systems of programs. When we learn that some relatively simple theory usefully describes a large set of phenomena, we bring it into the general network of scientific laws; similarly, when we find a specific application function that is useful in a large number of applications, we bring it into the network of systems programs. As time passes, the networks evolve into very useful, complex structures which no one man can hope to comprehend in detail. In programming this complexity has certainly been attained in the most recent generation of operating systems.

The number of instructions in large-scale operating systems has doubled about every three years to the point where larger systems have as much as two million lines of code. Operating systems of this size take on many characteristics of a human organization. No longer can the system be expected to perform exactly to specification at every point in its operation. Faults and idiosyncracies occur which cannot be predicted. The operating system, like the human system, must be designed to accommodate to these errors. It should not fail catastrophically; it should "fail softly"; that is, show only some degradation of performance. The most important change for the system constructor, therefore, is that he is no longer dealing with what to him is a simple monolithic deterministic system. He must use new principles of organization to construct the large systems.

The reasons for the growth in size and function are manifold, but the strongest ones arise from the trend toward multiple user data bases. The data base is no longer the property of a single individual. It is now used for communication between many departments and, perhaps, thousands of users. Extra care and control must be exercised to insure its integrity (correctness) and security (privacy). Since an installation cannot depend on each programmer taking the necessary precautions, these functions must be embedded in the operating system.

Another problem for the system constructor is the evolutionary nature of large systems. Human organizations are constantly changing in composition and direction and their change produces stresses for modification of the supporting information system. These stresses are so great that in many organizations more than one-half the programmers are patching up old parts of the system which perform no new function but which, nonetheless, need to be changed to perform correctly in the new environment. If we continued with our present techniques, the maintenance problem could place an upper limit on the size system we could construct. Fortunately, we are learning to construct systems which accommodate change more gracefully. Hopefully, this evolutionary capability will stand us in good stead for at least the near term future.

An even more important qualitative change is occurring in the systems program area. In the past, a user had to tell the computer exactly how to do his job. When the job is relatively complex, this is a frustrating, time-consuming task, particularly because most of the tasks are obviously straightforward to the human mind. In the information systems area, we have learned enough about the actions to be performed that we can now write systems programs which accept information from the user on what he wants done and proceed to determine how to do it themselves. This capability for accepting non-procedural instructions will continue to expand and make communicating with computers easier for both executive users and programmers. Since standard file handling processes account for about three-quarters of the application oriented code in use today, we have the means for solving the programmer shortage by using our best programmers to make the problems of the less capable programmers easier to solve.

In the application area, another major change has been the increased use of automatic sensors and testers. In IBM, for example, automatic testers are an integral part of the quality control-process control system. We shall see more of the impact of sensors in the discussion of our example systems.

Systems

At this point, let us consider two typical large-scale systems, one scientific and one commercial. It will become immediately apparent that cybernetic interplay between people of many disciplines and the computing system is at least as important a consideration in the success of the system as the details of system programming.

Earth Resources

Our present information about the utilization of earth resources even within the United States is more of a historical than real-time nature. There is an urgent need to improve this situation if we are to provide for more rational planning. One method of obtaining this information, which illustrates cybernetic aspects, is an information system to support data collection satellites which orbit the earth at several hundred miles altitude. There is a major cybernetic problem in combining all the resources necessary to produce an operational system. While the computer information system is to be central to the communications aspects of the problem, it is not the only portion that is in need of accommodation and development. Technical and human resources must also be developed. Means must be developed for obtaining indicative information from the satellite data. This must involve cross-disciplinary projects with members with meteorology, geology, agriculture, forestry, physics, chemistry and computer backgrounds. For example, measurements of crop blight would require meteorologists to provide the effect of cloud cover on the measurements, agriculture specialists to provide standards of measurement for blight and physicists to determine optical wavelengths for best discrimination, etc. In such a mission-oriented effort, all members must be able to work in compatible and cooperative terms.

In the study of a proposed system, it was found that the country was particularly lacking in human observers to interpret the full range of imagery from the satellite. There is, of course, a cybernetic trade off between the computer and the human observer with regard to sharing the interpretation load, but with existing techniques it is clear that we do not have nearly enough skilled interpreters to process imagery providing 100 foot resolution of all land surfaces every twenty days.

Clearly, in this environment of rapidly developing techniques and sensors, the computer information system cannot be the inflexible automaton with which we are familiar. It must evolve rapidly and easily to accommodate the human resources participating in the project.

In the commercial area, the need to maintain corporate information on a minute-to-minute basis is not only desirable, it has a clear economic payoff. The advantages of using resources such as airplane seats, manufacturing assembly lines, inventory or capital in an optimal fashion while avoiding resource consuming delays and mistakes are beginning to be quantified. Corporations are now developing large-scale information systems because they see near term payoff. Systems whose development costs range in the millions or tens of millions of dollars are expected to be paid off in two to three years.

A pioneer system in the commercial area is the SABRE system, a cooperative effort between American Airlines and the IBM Corporation. In 1957, the two corporations joined to construct a real-time airlines reservation system which would perform not just the seat availability task, but would instead handle the complete back room job of filing records, sending communication messages, processing waiting lists, preparing boarding lists, etc.

The system consisted of two IBM 7090 computers with sixty-four thousand words of two microsecond access time core storage backed up by one hundred million words of on-line disc storage. This system handles the 1,200 agents' terminal sets seen at American desks around the country. Few systems, even today, match the number of terminals supported by SABRE.

In the installation of such a pioneering system, there are many lessons to be learned. Among the most important things learned at SABRE was the need for condensed display of the real-time operation so that errors in programs interacting in real-time could be corrected. Without tools designed specifically for this purpose, it would have been completely impossible to track down errors caused by unique program interactions at the millisecond level. In addition, detailed measurements of performance of systems components were essential to provide a basis for system improvement. It is not possible to achieve optimum performance in a large-scale system *ab initio* simply because the system influences its environment and, in turn, the environment influences the way the system performs. Detailed analyses must be made of bottlenecks so that the system may be tuned to optimum performance. These analyses and measurements are early illustrations of the methodology for a new science of complex-interconnected systems.

New systems of even greater size and complexity are now under development and installation. It is not untypical to find systems having fifteen billion characters of storage on-line handling 3,000 messages per minute. These systems are even more complex because of the variety of information handled and the processing to which they are subjected. The system used by a manufacturer for handling his production control alone has interactions between seven separate subsystems located in more than a dozen plants in several countries. Systems of this size are a necessity for competitive large-scale corporations in today's environment.

The Human Organization

The human organization has not remained static in this rapidly changing environment. The most apparent changes are the organizational ones which have placed information systems leadership at the level of corporate vice president. Information systems are no longer fragmented and under the personal direction of middle-level management; their successful operation is crucial to the health of a modern corporation. Their implementation and operation are followed with active interest by top-level executives.

The crucial nature of the corporate information system becomes more clear when one looks at communication internal

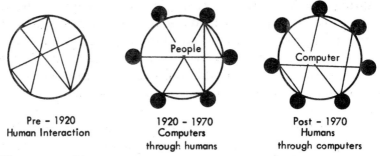

| Pre – 1920 | 1920 – 1970 | Post – 1970 |
| Human Interaction | Computers through humans | Humans through computers |

Figure 4. Phases of information system evolution.

to the corporation (Fig. 4). Internal memoranda will always exist, but in the transmission of business statistics in particular, a qualitative change has taken place. Initially, orders, accounts payable, accounts receivable, payroll and production scheduling were all maintained manually and results were communicated person to person. As time passed, humans became unable to handle the transaction processing efficiently and tabulating machines and computers were installed in various areas to process data locally. Communication of results from these local areas also involved considerable human effort in packaging, carrying, summarizing, etc. In these instances, the humans were an indispensable communications link between the local computers. In a large corporation, the requirements for collection, processing, summarization and dissemination of business statistics are so demanding in both time and volume that human beings can no longer adequately perform as communication links between local computers. In existing and planned systems, the computerized central information system will be a primary means of communication between people. Much of the statistical information will not be passed hand to hand for weeks through a distribution network; it will, instead, go immediately to the central computer files where it can be accessed immediately by all persons with a need to know.

Finally, the human organization for the creation of large scale information must often be dramatically different from earlier programming organizations. In many instances, the knowledge needed to create a system is dispersed over a hundred or more people. In such a situation, the design must be broken into smaller, comprehensible pieces with tightly specified external properties and interfaces. These pieces can then be given to groups of five to ten people for detailed design. Only in this way can the incentive for high quality detailed design be maintained. To a great degree, the large-scale human organization for creating large-scale information systems probably has the tightest policy and procedure control of any organization of its size to channel the great inherent flexibility of computer programs into a useful system.

Summary

The information collection and dissemination requirements of large, fast-moving projects can no longer be handled by unaided humans. The large-scale computerized systems which fulfill these requirements represent a quantum jump in information handling capability. They differ from traditional computer programs in several senses. No one person or small group can hope to comprehend and produce one million error-free instructions of program code. New systems must, therefore, be tolerant of error instead of failing catastrophically. Many more general functions must be included in multi-user data base systems to provide for security and integrity of the data, and the systems must be capable of easy accommodation to the normal rapid change of environment in the supported organization.

From the user's point of view the information systems will exhibit less need for step-by-step procedural instruction and instead, will show a slightly higher, but rigid, level of intelligence in being able to interpret instructions on *what to do*.

While we have mentioned the computer aspects in greater detail, the interplay between humans and large-scale information systems is of prime importance. If the computers cannot supplement the people in an acceptable fashion and the multi-disciplines of people cannot work together, then the most sophisticated program construction will be of little value.

Clearly, large-scale information systems provide an extremely important and fruitful field for future cybernetic study.

III
Direct Application
Of Cybernetic
Or System Techniques

Applications to Government

Science: The Good Urban Witch

ROBERT C. WOOD

Under Secretary, Department of Housing
and Urban Development
Washington, D.C.

I wish to direct the potential contribution of science and scientists to the most difficult problems of cities and urban development. My enthusiasm for a marriage between science and urban affairs goes back a number of years and I'm sure that some of my associates in Congress and the housing industry, not to mention certain academic colleagues, regard it as excessive or even peculiar.

The consumation of such a union now seems likely—but the timing is ironic. Urban policy, after a prolonged infancy, has come of age. The 1968 Housing Act is the most far-reaching and sophisticated piece of urban legislation this country has known. It attaches numbers and a timetable to the achievement of our long-standing national goal of "a decent home in a decent living environment for every American family."

Our need for science, for innovation and ingenuity, is great. We are ready. But at just this time we are seeing a noticeable change in the temper of the public regarding the ability of science to resolve national problems. This change is evident in the relative parsimony of that sensitive barometer, the United States Congress. It is evident in the anti-intellectual

123

asides of the third-party candidate; it is evident in the sharp questions of the young and the poor.

This lack of phase is the background for my rather straight-forward thesis this evening. The proposition I would put before you is this: The value of systematic scientific research on urban problems cannot be assumed to be self-evident. It must be demonstrated by our progress not only on limited, obviously technological questions, but on the more intractable ones.

A couple of months ago, a key HUD staff member sat in on a HUD-supported summer study on "citizen participation" in various aspects of city administration and planning. When the subject turned to health care, someone suggested that the use of systems analysis could be extremely valuable in improving the delivery of health services in the city.

A Negro health official responded with some heat. "I'll tell you what the community will say to that. They'll say systems analysis is just another part of the white establishment. We don't want to hear about that stuff. The whites have been doing a miserable job. Infant mortality in Harlem is forty-two per 1000. Give the community control of the health program. Period!" After the meeting, my colleague went up to this man and asked him what would happen if after three years of community control, the infant mortality rate was still forty-two per 1000. "Then," he replied, "we'll use systems analysis."

One of the most difficult problems we will face during this decade is the transfer of management capability to new groups within the population. The above is an illustration, as is New York's effort at school decentralization.

It seems clear that the neighborhood has become the key unit in social programs. To rationalize this with the complex interdependence of our metropolitan and interstate regions is going to strain our administrative as well as our research capacity, but if we are to govern urban America, I believe we need a greater consolidation of authority at the metropolitan level and a decentralization of authority at the neighborhood level.

America's great urban regions now lack the powers to guide

the course of their development. They cannot decide the use of their most precious commodity—open land—nor prevent the fouling of the air and water, nor assure equality in education and opportunity for their children. Until they have such authority, until suburb and central city acknowledge in these specific respects their common concerns, we can blanket the present array of local jurisdictions in a blizzard of Federal cash and still fail to protect our urban heritage and upgrade our urban environment.

At the same time, however, as our regions grow larger, as the typical American urban community numbers its citizens by the hundreds of thousands, we need to put certain powers closer to the people. We need to decentralize both debate and action on those community activities that affect intimately the everyday life of every citizen: housing, schools, health care, jobs, public protection. The restoration of the neighborhood as a vital place whose residents have a sense of purpose and belonging is as great a challenge of the American system today as regional development. Only in the neighborhood can the appearance and pace of life be changed.

Citizen participation is "what's happening" today in our cities and it introduces pretty volatile variables into some of the urban sub-systems: education, planning, housing, health, and police protection—among others.

The issue before us is not citizen participation as such but the loss of credibility. If you talk to the articulate students on our campuses today, or to the ghetto residents, you find a loss of faith in rational approaches to our urban dilemmas. They are turned-off by talk of relative gains or incremental improvements. They are interested in absolutes, in values, in total change.

It is easy when faced with this new mysticism to retreat behind the generation gap. There is a fair amount of youthful provincialism in the New Left's visceral indictments. And even college professors sometimes react to it with middle-aged impatience.

But if we refuse to listen to the student skeptics and the black community, we do so at our peril.

Just a few years ago, America's vast technical and scientific capacity was clearly cast as the Good Witch of the South. Now, to many Americans it's the Wicked Witch of the West. The same could be said of computers and automation.

This change cannot be attributed solely to the Vietnam war. R & D are the victims of oversell as well as overkill. The technological fall-out from the space exploration program that was so freely predicted, for example, hasn't materialized. Twenty-five billion tax dollars have gone to NASA, says Mr. Average Alexandrian, and it still takes me an hour to get across Memorial Bridge.

We can keep Schirra and company healthy in the hostile void of outer space—but we can't keep babies from dying in the ghetto.

In the wake of Sputnik, spending on educational research shot up. Local school districts scrimped and taxed to buy the latest textbooks and language labs. Per pupil expenditures increased fifty per cent in ten years, not counting inflation. While this adventure brought imaginative changes to the teaching of science, it did not really touch the hard core problems of urban education. The much-advertised harnessing of technology to the science of learning has so far produced some ingenious teaching machines, but almost no decent software.

You and I both know that educational research has not been notably rigorous and that the Apollo will ultimately produce dividends. We know the folly of what John Gardner used to call "the vending machine concept of social change"—put in a nickel and out comes a candy bar; pass a law, appropriate some research funds, and out comes a solution to a social dilemma. But we are dealing with a younger generation that accepts no excuses and with a black community that has grown tired—after a hundred years—of waiting on the doorstep.

What are we to say to *them?*

For several years now I have been excited by the possibilities that lie along what the Defense Department might call the science-city interface. I still believe that the creation of a viable urban future will depend on just that frontier.

In 1966 we held a summer study together with the President's Office of Science and Technology on "Science and the City." Biologists, economists, physicists, sociologists, housing specialists and others met together for two weeks in what, these days, might be called a confrontation. The ripples of that experiment are still rocking the boat at HUD, in the best sense.

Also in 1966, Congress was persuaded to give HUD funds for an Office of Urban Technology and Research. The first Director of that Office, Tom Rogers, came aboard to handle the first general research appropriation of $500,000—an amount, it should be said, that was less than half the annual amount available to the Department of Political Science at MIT. The FY 1969 appropriation will be $11 million—a 22-fold increase from our initial allotment and a ten per cent increase over 1968. Given the nature of this year's general R & D appropriations, this is something of a minor miracle.

Last spring, the Urban Institute was founded to support research with a combination of public and private funds. The caliber of the Board and of the staff already assembled by President William Gorham offers considerable hope for the future.

Urban research is a growth industry. And it had better grow fast. It is becoming increasingly obvious that we know painfully little about cities and about the long-term effects of them on our present patchwork of urban programs. In a *Fortune* article last summer on the welfare program, Edmund Faltenmayer wrote:

> Belatedly, the country is beginning to search for a better way to do the job. The growing clamor for change, however, has been accompanied by an almost total lack of public understanding of what a welfare program should be and do. Because of a scandalous lack of research, the public—and this includes Congress—has been advocating policies and passing laws on the basis of *beliefs* rather than facts.[1]

HUD's research money, initially, is being used in two ways: first, in trying to learn how to conduct experiments in the city; and second, in looking for ways to loosen up urban institutions so that they can respond to new ideas and accept rational solutions.

The first of these tasks deeply challenges the hard sciences. We have got to find ways to conduct decent experiments in situations of extreme sensitivity in which we face:

— the full blare of public attention and publicity
— a good deal of nonprofessional advice that is also politically powerful
— an inability to control easily or isolate the elements with which we wish to deal
— conclusions that will never be sharp and unambiguous.

For scientists brought up in a world of laboratory experiments, of protected privacy, and of results which are judged by professional colleagues, this is a rough transition. But it must be made if urban America is not to be condemned to ever-more-serious repetitions of her past mistakes.

A major HUD research effort is now underway. The pioneering "in-cities" project will attempt to specify, in the engineering sense, what the constraints are that inhibit reductions in cost and time in building homes for low-income families *where they need them.*

We do *not* expect this project to develop any new technology nor many approaches which have not been suggested before. We *do* expect it to develop reliable, quantitative information about the effects of the more important constraints which inhibit the introduction of innovation, or reduce the effectiveness of such innovation. We *do* expect that this information will be developed in such a way that it can be confidently extended to large-scale applications, and to innovations which may be developed later. We *do* expect to develop it under dynamic experimental conditions where we will learn what the true advantages and limitations of various innovations are, and what the constraints actually are—rather than what they are imagined to be when studied historically or analytically. And, we *do* expect to record this information carefully, comprehensively, and objectively, and to disseminate it throughout the nation, promptly and publicly, making it available to all who bear important responsibility for housing our lower-income families.

The effort is politically sensitive, concerning as it does codes, zoning, Federal and municipal red-tape, union regulations, and integration. But, as my technical tigers say, "We're going to run right smack into the sausage machine. We're going to *measure* and we're going to publish."

In this context of dealing with the important and the almost intractable the unique analytical capability of the cybernetics fraternity, the knowledge of control and communications and information systems, is a considerable national resource and one which should be coupled to our urgent need to know the American city and to keep it manageable, to restore the options and the beauty to urban life.

The kind of analysis we need, however, is not that which begins "I imagine the city to be thus-and-so." We've got to deal with the problems that grip people *now*. We cannot assume a clean slate. Manhattan is not going to be returned to the Iroquois. If, on the other hand, you are ready to undertake urban analysis that starts from a good theoretical base and uses a broad foundation of data—begin at once. You will find, of course, that neither theory nor data are at hand and your urgent first task is to develop them.

William H. Whyte, the author of the Urban Beautification program among other things, writes tellingly about the affection of planners for the clean slate and the grand design:

> To arrive at [a grand design], the planners hypothesize a number of alternatives. The first will be termed "unplanned growth" or "semi-planned growth" or "planned sprawl." Actually, this will be the most challenging alternative to work with, since there is a fair chance something like it will come about. It is anathema to planners, however, some of whom would as soon go to hell with a comprehensive plan as heaven without one. They go on to sketch the grand designs, such as rings of satellite cities, radial corridors, or wedges. After the pros and cons of each are matched against the others, the optimum is chosen. Whatever its geometrics, it will call for a sweeping rearrangement of the region with its growth dispersed into self-contained new cities, each separate from the other by huge expanses of open space.
>
> Designs can indeed help shape growth, but only when the designs and growth are going in the same direction. Most of the year 2000 plans are essentially centrifugal—that is, they would push everything outward away from the city, decentralize its functions, and reduce

densities by spreading the population over a much greater land area. I think the evidence is staring us in the face that the basic growth trends are in the other direction; that they are toward greater centralization and toward higher rather than lower density.[2]

There is a great deal we don't know about density, about the second order consequences of packing huge numbers of people into a limited space. We don't even know, as Mr. Whyte points out, whether high density is a bad thing, although some provocative insights have been put forth from observations on animal behavior.

As we try to confront these and other difficult questions, we of the urban fraternity desperately need your help. But in arranging this scenario, as I indicated earlier, some of our most important critics are in the Congress.

The *relevance* of science and research must be established anew with a skeptical Congress and with a generation that wasn't around when radar saved England from the Nazis.

We cannot dismiss the concerns expressed by Senator Gordon Allott of Colorado in his letter to *Science* magazine:

> For some time, I have been warning members of the scientific community that unless some adequate means are developed so that the taxpayers and their elected representatives know what they are 'buying' with their research dollars, a reaction would set in one day which would cause a severe cutback in funds allocated for research. . . . With so many other problems facing us today which require immediate attention and the expenditure of huge sums of money, it is a bit unrealistic to ask the taxpayer to continue to finance and support an annual $17 billion or more R & D budget almost on 'faith' alone. . . .
>
> The admitted lack of expertise on the part of a majority of members of Congress in areas relating to scientific achievements is, regrettably, matched only by the lack of appreciation on the part of many research scientists, engineers, and technical managers of congressional processes and problems. . . . Gentlemen, we have only so much money to expend. Within our admitted lack of expertise, coupled with an appalling lack of national goals or systems of priorities, I think we do a fair job of spreading out the federal dollar. We could do better, though, with some constructive help from the scientific community from an objective and realistic appraisal of the circumstances and of existing realities. I would think that the country might well benefit if, paraphrasing both Donald Hornig and the 'now' generation, the scientific community would become 'in-

volved,' would drop the cloak of mystery, and take the time to explain, not just to us in the Congress, but to Mr. Taxpayer as well, just what it's all about. . . .[3]

It is HUD's intention that the relevance of urban research be firmly established—but not by conjuring up the successes of military research that might or might not be analogous; and not by playing unfairly on the fear of urban turmoil.

Together we can establish that rationality and empiricism are still the best way to approach the discontinuities in our environment—and to do so in tangible, visible ways that lead to better conditions for our people.

REFERENCES

1. Edmund Faltenmayer, "A Way Out of the Welfare Mess." *Fortune, 78,* No. 1, July 1968, p. 62. Courtesy of *Fortune* Magazine.
2. William H. Whyte, *The Last Landscape* (New York: Doubleday & Company, Inc.), Copright © 1968 William H. Whyte. Reprinted by permission of Doubleday & Company, Inc.
3. Gordon Allott in a letter to *Science, 162,* October 11, 1968, pp. 214-218. Copyright 1968 by the American Association for the Advancement of Science.

City Halls and Cybernetics

E. S. SAVAS

Deputy City Administrator
New York, New York

Introduction

A Greek word that's known to all of you is κυβερνητης. When transliterated and anglicized, it has provided the name of this distinguished society. In modern Greek, that word means governor, that is, the chief executive of a government. Linguistically, therefore, there is an equivalence between the chief executive of a government and the cyberneticist. What I would like to do today is explore that similarity in a big-city setting.

First, let us look at the simple feedback control system of Fig. 1, where the desired value is fed into a comparator, and any discrepancy between desired and measured values results in control action which tends to reduce the discrepancy. The system is designed to overcome the effects of the external disturbances which impinge on the system.

In Fig. 2 we see the governmental analogy to a feedback control system. The goal-setting mechanism, which establishes objectives for the system, can be viewed as the output of a planning, programming, and budgeting system (PPBS). It results in set points for the city.

The process of taking control action to reach the objectives is the process of administration; that is, the bureaucratic pro-

133

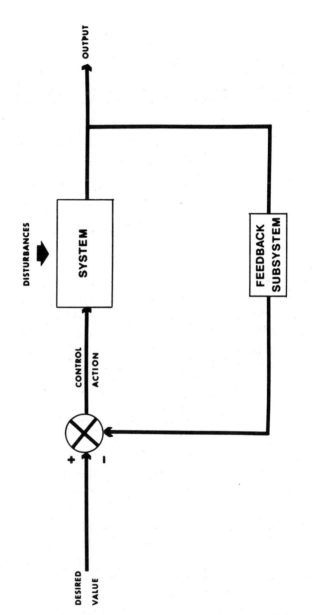

Figure 1. A conventional feedback control system diagram.

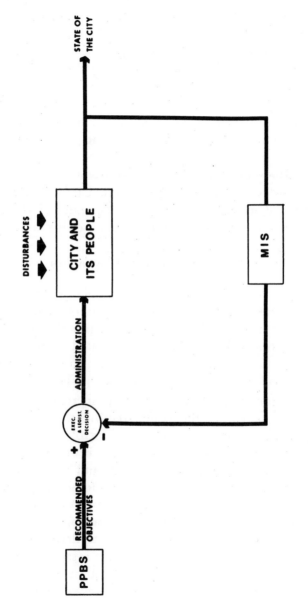

Figure 2.   City government as a cybernetic system.

cesses of the city can be viewed as the actuator element, strange as it may seem. The system being acted upon is the city and its people, which are subject to disturbances that are social, economic, political, and natural. The output of this living system can be defined as the state of the city. Feedback from that state, the measured condition of the city, is provided by a management information system (which may or not involve some computerization) and fed back into the comparator, which produces executive and legislative decisions.

Feedforward control can be applied, of course, to the extent that disturbances can be anticipated and accommodated within the planning portion of the planning-programming-budgeting system.

In the remainder of this paper I would like to discuss some recent developments in New York City in five of the areas indicated in Fig. 2: the overall dynamics of the process, the feedback system, administration, goal-setting, and disturbances.

Governmental Dynamics

The concepts of industrial dynamics apply also to the governmental dynamics implied in the control loop of Fig. 2. The first disturbing thought that crops up is that the natural time constants of urban systems are unrelated to the term of elected office, and that means that there is a potential for some instability and erratic *hunting* of the set point.

The long-time constants and the incredibly involved multi-variable nature of the system require that a very large, very sophisticated, very complicated control device be employed, a device we call government. But it is difficult to keep a large, sophisticated controller tuned up, and there are always component failures, slippage of gears, loose connections, etc. A conventional control-engineering approach to this kind of a problem is to apply minor-loop control and cascade control, although we recognize that complete decoupling of variables cannot be accomplished and that one is sacrificing the optimum that *theoretically* could be attained by the more integrated total system. Decentralization is an example of minor-loop control. As you know, we have been having some initial troubles in New York City with decentralization as applied to the school system—getting a new controller on-line is always troublesome. Nevertheless, both from the standpoint of cybernetic theory and from the standpoint of good government, decentralization does make sense. Getting decision-making down into the community offers hope of getting more rapid and more effective response of the system to achieve its performance objectives. So, the concept of school decentralization, and of participatory democracy in general, is in accordance with cybernetic principles.

In addition to the very difficult step of implementing decentralization, another step being taken by the current administration to improve governmental dynamics and to provide participatory democracy is through the activation of community planning districts. Neighborhoods which are historical and topographical entities have been formally recognized as community planning districts. They are receiving a modest amount of staff support, a modest budget with which to determine community interests and a mechanism for expressing this interest that goes far beyond the conventional opportunity to write letters to their elected officials.

The concept of community corporations, funded by the Office of Economic Opportunity, is yet another approach being implemented in New York City to improve the performance of our complex system.

Feedback

Let's turn now to the second of the five areas, the feedback subsystem, and let us begin by examining a mayor's sources of information and the characteristics of those sources, as a communication engineer might look at them.

One source that a mayor has is direct visual observation: he sees things as he goes about the city. Certainly, this is a good information channel, but it has only a limited band-width signal and very limited line capacity.

A second source of mayoral input for decision-making is

the information provided to him by his subordinates. This is a sound, reasonable, acceptable, and good feedback source for the mayor, but there are problems that should be recognized with this kind of signal as well. There is selective amplification of the signal, and a mayor, like any chief executive anywhere, must ask himself, "Why is this particular individual giving me this information at this time?" It could be, for instance, that it is an effort to get a larger piece of a limited budget, or it could be an attempt to whitewash an incipient problem in the subordinate's area of responsibility.

The press represents yet another information source for a mayor—a good source but one that issues only carefully filtered signals that emphasize dramatic events.

A fourth valuable source of information for a mayor is the *establishment,* the leaders of political, religious, business, labor, and academic groups, for instance. One must recognize, however, that these are high impedance channels; that is, the signals that they issue may generate a self-induced resistance and, therefore, careful matching throughout the entire communication circuit is necessary in order to conduct the signal faithfully from the individual members of the group to the mayor without excessive distortion.

Other channels of information to the mayor come from other public officials at other levels and from other branches of government. These may also suffer from problems of high impedance.

The final source of mayoral information identified here is the public at large. Four particular subgroups within this source are worth considering. One consists of highly vocal and vociferous individuals. The problem with their signals, of course, is that they have a rather low signal-to-noise ratio and, therefore, may require long-term averaging in order to extract meaningful information from them.

On the other hand, special interest groups emit signals which have very high signal-to-noise ratios, but are biased signals.

Elections represent the classical democratic approach to information feedback for the mayor. The problem with elections, however, is that they are characterized by very low sampling rates (once per four years) and they produce only one bit of information: go or no-go!

Finally, in these times of stress, civil disorders have to be viewed as information-bearing signals. Unfortunately, these signals can be rather powerful; they saturate the system, which tends to trip the circuit protectors, with the result that the only information received is that the system has failed.

Given these characteristics of the feedback process, one can identify ways of improving the flow of usable information to the mayor, and cite some recent examples of improvements in New York, improvements which increase the sampling rate, provide more feedback channels, increase the band width, enhance weak signals, suppress noise, and correct biased signals.

One example was to bring government from City Hall into the community, out of downtown and into all sections of town, by opening up little City Halls throughout the city and sending mobile City Halls to set up shop temporarily in various neighborhoods and find out what people are thinking about, what the problems are in that specific neighborhood, and what the problems at large are that are bothering the people.

A second example of improved feedback is Mayor Lindsay's Urban Action Task Force, now being emulated in other cities. In essence, the objective of the Task Force is to bridge the gap between the alienated people of the city and the city government. A number of deputy commissioners, assistants, and others in the government have been assigned liaison roles, in addition to their normal duties, with various community groups. They tune in on the same wave length with the community and keep in touch with local community leaders, youth groups, and community corporations.

The Task Force is significant because people who never before recognized that there was any such thing as government that was relevant to them find, in fact, that they too have access to the power centers in government and that there is someone listening to them who can often get things done for them, whether it be building "vest-pocket" parks or getting the streets cleaned. The Urban Task Force operates through high-level coordination with each of the city departments, so that, for example, the Sanitation Department can be called in to clean up garbage-strewn lots; the Highway Department can come in and black top them; and the Parks Department can then set up a basketball hoop. It's hardly the solution to America's urban ills, but in terms of short-term action, it seems to be effective in bringing people into the political process and into the governmental framework of problem solving.

A third aspect of improving the feedback mechanism in the city is through the Mayor's Action Center. This is simply a round-the-clock operation where citizens can register complaints. The Action Center handles 200,000 telephone calls and 5,000 letters each year. It is manned in part by volunteers and, as with the Urban Action Task Force, operates through high-level, departmental liaison personnel. Complaints are handled as high priority items by the department and ten-day response is guaranteed.

A fourth new element in the improved feedback system is the night mayor. High-ranking city officials are on a roster and each periodically serves as night mayor, stationed overnight in the mayor's quarters at City Hall. He answers urgent phone calls and acts in case there are serious incidents. It is important that someone is minding the store in the middle of the night.

Another important feedback channel has recently been opened up, this one to the university community. The first significant step was the appointment of Dr. Timothy W. Costello, a professor at New York University's Graduate School of Business Administration, as Deputy Mayor-City Administrator. Next, the Mayor started holding some of his regular cabinet meetings on the campuses of the seven Ph.D.-granting institutions in the city, providing an opportunity for the academicians to see what city government is like, and also exposing city commissioners to the activities in the various universities that may be of relevance to them.

More recently, the city has established an Office of University Relations to encourage and facilitate contacts between universities and the city government. In another vein, but still dealing with university feedback, New York City was the first to pioneer the Urban Corps, which is a summer intern program. Last summer we had close to 3,000 college students from about 100 colleges and universities throughout the country who worked as interns in various city departments. It proved to be a fruitful experience for the students and a breath of fresh air for some of the mustier regions of our inherited bureaucracy. The city has now received a grant to spread the gospel and show other cities how to set up and operate Urban Corps programs.

Computer-based management information systems represent the glamorous, technological approach to improved feedback, and New York City, like other large organizations, has begun to move in this direction. Like many private organizations, the city in the past has emphasized the clerical aspect of computer usage. We now recognize that the main value of computers is to use them more effectively to enlighten the decision-making process. That's not so easy to accomplish. There are some significant problems in the way the city organizes, manages, and uses its computing resources, and the changes that are needed are very fundamental management changes. We have under way a major project to tackle these problems and to produce a master plan which will enable the city government to take full advantage of modern information technology. This involves changing the way we make decisions about how to buy computers and what to use them for.

Administration

Now, let us turn to the third of the elements identified in Fig. 2, the control subsystem, the administration of the city, the bureaucratic apparatus, if you will. One of the most significant and important but undramatic accomplishments is a thoroughgoing overhaul and reorganization of the city structure, consolidating forty-nine separate departments into ten superagencies. As an example, one of the superagencies is the Environmental Protection Administration, which consolidates the Department of Sanitation, Department of Air Pollution Control, the Department of Water Supply, and the Bureau of Waste Treatment. In other words, liquid, solid and gaseous wastes are being treated as an entity and, hopefully, we will evolve more coordinated and comprehensive policies of waste management.

A second major administrative improvement is the long-awaited revision of the building code. This sounds terribly prosaic, but New York City had a building code which was originally written in the 1890's and modified extensively in 1937. The new performance-oriented building code will have a significant impact on the cost of building in New York City, and will enable builders and architects to capitalize on new techniques and new materials.

Now let us review a number of recent improvements and pending improvements in administration that are based on advanced usage of computers. First, let me describe one of our pet projects, GIST, the Geographic Information System. It is addressed to the interrelatedness of urban problems and focuses on the fact that numerous city agencies are concerned about buildings and property: where it is located; how much it's worth; who owns it; what it's used for; what's built on it; what is its condition, its rentals, its tax status; whether there are inflammable materials stored in it; has the boiler been inspected recently; does it have an incinerator and has the incinerator been upgraded; etc. The result is fragmented record-keeping, which makes it very difficult to exchange related data among departments. An apocryphal story will illustrate the point. A fire occurred in a certain city and the building burned down. The fire marshal was trying to find out what caused the fire, and the first thing he did was to go to the boiler inspection unit. "There was a fire at 125 South Eighth Street," he said. "When was the boiler last inspected?" The boiler inspector responded, "I don't know, what's the number on the boiler?" And the fire marshal said, "I don't know what the number on the boiler is, I'm telling you what the address is." "Sorry," was the rejoinder, "our information is filed by boiler number."

Now the problem indicated here is not an intellectually demanding one. Conceptually, it is trivial and could very easily be glossed over in any elegant presentation about feedback and cybernetics in cities. However, it just so happens that these are the problems that one really stumbles across when one is on the inside trying to solve urban problems. Well, we think we have an approach with our Geographic Information System, which essentially will permit departments to exchange property-related information regardless of which of several different locational identifiers is used: street address, tax block number, census block number, health district number, police precinct, etc. Not only will one department be able to pick out vitally needed information from another's files, but it will also be able to prepare computer-generated maps which will show at a glance, for example, the geographic distribution of such widely diverse factors of interest as tax-delinquent properties, ambulance calls, dilapidated buildings, warehouse space, burglaries, or school-age children.

A very different example of improved administration in the cybernetic loop is a computer-based command-and-control system for the Police Department, which will reduce the time it takes for police cars to reach the scene of a crime. It will work in the following way. A person dials the police emergency number (911). A policeman in the centralized Communications Bureau answers it and receives the information. He makes a few insertions on a keyboard in front of him, which enters into a real-time computer the location where a police car is needed. The computer then scans its file and selects the nearest available police car and displays its call number on a screen in front of the radio dispatcher. The latter radios that car and instructs it to proceed to the desired address. This will cut down the response time of police vehicles and at the same time serve as a convenient means of capturing source data for subsequent analysis.

To continue with this list of improved administration via computers, our Manpower and Career Development Agency is using a computer for job matching. The computer belongs to the Bedford-Stuyvesant Community Corporation, which provides this service to the city. Jobs are entered into the computer as they become available. Job applicants in underprivileged areas of the city can go to any of fourteen neighborhood Manpower Centers where a counselor interviews them and enters information about the applicant into a console which is connected by telephone to the remote computer. The computer examines its job file and reports back immediately if a match can be made.

In the field of education, the cybernetic principle of reinforcing behavior is being applied by our Board of Education through an experimental system for computer-based instruction. More than 6,000 students in sixteen schools use the system for drill in reading and arithmetic.

The Department of Traffic Control will soon activate a computerized traffic control system to control the signal lights at 500 intersections. That is the first step, which, if successful, will be extended to about ninety per cent of the city's traffic signals.

Goal-setting

Now, let's turn to the fourth major element, the goal-setting subsystem. What are the recent improvements in New York City in terms of goal-setting, in setting objectives for the city? Well, let me toot our horn here a little bit. New York City was the first city in the country to establish a Management Sciences Unit; the unit is headed by a Deputy City Administrator (myself) who reports to the Deputy Mayor-City Administrator, Dr. Timothy W. Costello. It represents an effort to build up within the city the internal capability to utilize the tools of systems analysis and operations research, tools which can be used to identify and quantify government goals.

One of our first major outputs was a classical quantitative study, replete with computer simulation, mathematical models, and cost-effectiveness analysis, of the city's emergency ambulance service. We were concerned with the slow response of ambulances and turned to computer simulation as a useful tool in examining the factors that affect response time. We established the goal of reducing response time. Our study determined that the least expensive way to achieve a substantial improvement was to reorganize the service, remove ambulances from the control of individual hospitals, and redeploy them broadly in accordance with the observed demand. We are pleased that the United States Department of Transportation has recognized the significance of this work and has provided us with a grant to extend it further and to make it available to other urban areas as well.

We also established an objective to reduce voting delays during elections. To do this, we used computers on a trial basis to form more equitable election-district boundaries and thereby achieve a better distribution of our voting machines with respect to population. Incidentally, the kind of computer-based capability that is implied in an election districting system would be part and parcel of GIST, described above, because it requires the ability to manipulate geographically-oriented data.

A most important innovation in the city's goal-setting mechanism is the introduction of a planning-programming-budgeting system. Right now, all our departments are deeply immersed in the thorny problems of learning what PPBS is, converting their traditional line-item budgets into program budgets, and trying consciously to relate budgets, programs, and departmental objectives.

Disturbances

Let me turn now to the last element, the disturbances which affect the system. Disturbances are the independent variables which act upon the system from the outside and over which local government has no direct control. Disturbances are social, economic, political, and natural. As an example of a social disturbance, one can cite the revolution of rising expectations that is affecting not only our cities but the entire world. Recessions and wars, which have a cataclysmic effect on the economy of the city, illustrate the economic disturbances which are beyond the control of a mayor. Political disturbances, which have a profound impact on urban programs, affect the city when changes in administration occur at the

state or national level. And, of course, there are natural
disturbances, such as the droughts, floods, hurricanes, and
earthquakes that afflict some cities.

Let me cite some specific examples of government-induced
disturbances, which originate at other levels of government.
It is quite clear, for instance, that the welfare policies of the
nation as a whole and as implemented in certain states influence the rate of migration from rural shacks to urban slums.
As another example, a recent change in federal law led to a
change in state law which forced New York City to drop forty-
three per cent of the participants who had been enrolled in
the Medicaid Program. That represents over one million
people. This large and sudden load fluctuation causes havoc
with the administrative end of the procedure and results in
the appearance that cities are bungling and incapable of
handling routine administrative matters. Furthermore, the
cutback in the Medicaid program will boost tremendously the
number of patients who come to the municipal hospitals for
service.

Conclusion

Let me close now by saying that we can indeed look at our
municipal governments in terms of the basic principles of
cybernetics. In New York City we have a number of activities
underway that give us confidence that the tools and concepts
of this discipline can be usefully applied to urban problems.

Systems Engineering and Urban Problems

EDWARD H. ERATH

CHARLES DRESCHER

Los Angeles Technical Services Corporation
Los Angeles, California

We are certainly living in interesting times. Our technological progress has far outdistanced our ability to cope with or effectively utilize it.

It is a time when man seems to be able to invent whatever is required to achieve some of his age old dreams. Traveling to celestial objects is a startling example; but what about our interest, or ability, to deal with down to earth problems? The agony of our cities.

This is a time when everybody seems to be talking about the problems of our cities; and they should be. But not enough is being done about them. Our cities are in deep trouble; the problems are complex, probably the most critical yet faced by civilization, and they seem to expand and intensify faster than solutions for them can be identified and brought to fruition.

All of our institutions seem to be under attack, confronted by problems of growth and resulting complexity for which no experience or training seems to have prepared us. What are

147

we to do? Some say build new cities; others say build new cities inside of our old cities; but, the fact remains that we must learn how to rejuvenate and live in our existing cities—a task that seems to paralyze us.

Where can a start be made? How can progress be implemented to achieve significant results rapidly? What experience or capability can we call upon? Fortunately, successes attained in research and development programs of the Federal Government in space and defense offer hope. These programs have been marked by remarkable advances in technology and management techniques; their hallmark is the application of diverse and extensive resources to massive tasks. System analysis and system management are the tools used, in fact, the tools invented to meet the needs of national security. These tools and this experience can be focused on defining the problems inherent in making cities more desirable places in which to live and work. They can be used to invent whatever is necessary to achieve this objective. However, to say these tools are applicable is one thing; to apply them is slightly more difficult. There is a great lack of understanding about their nature, potential, and limitations in non-defense spheres; even managers and technologists in industry don't easily recognize their transferability to other, non-defense, non-space tasks; local government officials are reluctant to utilize them; in some cases, they have been oversold.

Recently, to demonstrate their applicability and that of the technology they have spawned, the Los Angeles Technical Services Corporation, with the cooperation of the Air Force, arranged a visit to North American Air Defense Command headquarters to demonstrate for a group of skeptical city officials and business leaders the status and deployment of Forces techniques, information handling, and communication systems used by the Air Force, and to show how these techniques were applicable to city police and fire department operations. We were all surprised—the skeptics by what they saw, our technologists by a statement made by Mr. J. R. Lakey, an Air Force civilian scientist:

> It never occurred to us until you people contacted us, but what we do to defend the continent and what a police department does in the city are remarkably similar in principle. The weapons used are different, the motives of the aggressors are different, but the underlying operational principles and requirements are identical.

Little wonder there are skeptics, when the specialists don't readily see the potential.

Another impediment to rapid adaptation of these tools to urban problems is the lack of clear-cut city objectives or goals. Urban goals seem nebulous when compared to national defense objectives. If a national objective is to maintain the North Atlantic Treaty Organization, NATO, one military mission is to maintain open sea lanes across the North Atlantic. Defining this mission identifies many military requirements, or weapon system potentials: fast cargo ships, fast escort ships, antisubmarine airplanes and other aircraft, antisubmarine warfare techniques, and a host of other alternatives. This is where system analysis techniques are brought to bear. Within achievement of the military mission, system analysts determine the combination or mix of weapon systems that best meets mission requirements, consistent with the nation's financial ability. This is a critical phase in the national defense program, since the nation cannot afford to buy all of the systems proposed to satisfy a given mission, and yet national survival must be insured by the choices made.

No such process takes place in local government. No mission analysis is performed, procurement judgments are usually too subjective and often result in an unnecessary procurement or serious mismatch between actual operating requirements and what is available off-the-shelf. Certainly, city government with its limited financial resources should be, if anything, more concerned than the military about the appropriateness of its procurements. Consequently, it should be at least as essential to city administrators to define carefully their objectives and to analyze their requirements prior to any procurement program, to avoid the possibility of dissipating their meager financial resources and to avoid the equally undesirable consequences of failing to respond to the needs of its citizenry.

We often hear local leaders say that if we can fly man to the moon, we should be able to get him to work on time. The keys to achieving the former are a unitary objective, a single responsible agency, a wealth of professional talent, and money. The Apollo Project requires a continuity of effort and singularity of purpose over at least a decade; a single agency competent in managing diverse resources and responsible for achievement of the objective and sufficient money, at least $20 billion, to support the efforts of 20,000 contractors and 300,000 workers in universities, government agencies, and industry.

Getting man to work on time in the urban environment is a much more complex and difficult task; but where is such a combination of resources at work on the problem of mass transportation? Or, for that matter, on any other important urban problem? If industry and the scientific community are to undertake a meaningful, systematic approach to solving the nation's major domestic problems, those of the cities, appropriate arrangements will have to be made to mobilize their talent on a scale at least equal to that required for space flight. Even then, without appropriate action within local government, progress will be difficult to achieve.

What kind of action can we take? As early as 1963, far-sighted city officials in Los Angeles were taking action to correct that City's deficiencies by taking advantage of existing modern techniques—to adapt the City's operations to available tools, and vice versa. In December of that year, the Los Angeles Data Services Bureau was formed by executive directive and council ordinance. Prior to this time, twelve city departments had their own data processing facilities. Each department proceeded with its own installation, often unaware of what was being done in other departments. Consequently, there was considerable overlap of effort and duplication of costs. The formation of the Data Services Bureau as a centralized service facility, the first of its kind in any major city, gave Los Angeles a running start in the utilization of computers to solve complex urban problems. The Bureau's operation has resulted in proportional reductions in expenditures and is considered an unqualified success, even by individuals originally opposed to its establishment.

A similar and more ambitious effort had been initiated earlier in 1963. As in the pre-Data Services Bureau days, communication systems, equipment, and services were being designed, procured, and installed by many different departments, independently of one another. So the City consolidated those services in one department. However, the problems involved in centralizing communications are much more complex and many-faceted than those of data processing. Consequently, the City's communication and command-control operations for emergency services, like those of all cities, are far behind state-of-the-art technology.

Recognizing that even more powerful system development tools existed and should be brought to bear on the city's problems, the City, with the mayor's leadership and the council's support, created its own non-profit system engineering company, the Los Angeles Technical Services Corporation, and asked it, among other things, to begin development of modern communication facilities for the City's departments. All of these actions are indicative of the innovative steps local government can take to improve its operational capability and to exploit modern technological developments.

The Corporation received a voluntary contribution from a large aerospace company and subsequently funded a program to review all of the city's communication systems, techniques, and plans for future development.

It did not take long to produce an obvious result. After a month of effort on the program, we were not too surprised to receive an unscheduled visit from the program manager, who said, "We are deeply concerned over an almost total lack of comprehension in city government of the potential for improvement in operations that modern technology offers, a lack of awareness of the cost and performance benefits that can be achieved, and the absence of experience necessary to manage complex development programs." City officials, by contrast, say that industry specialists are too "blue sky." Obviously, more must be done to bridge the gap. Yet, what to do is not obvious!

Perhaps one engineer's comment in a trip report to City Hall best portrays the underlying problem involved in matching industry specialists and their skills to the needs of the cities. The engineer had just toured the City's communication facilities and remarked:

> Our trip through the dispatch and emergency centers vividly confirmed the obsoleteness of the bulk of the City's communication equipment and procedures. Their mobile communications center reminded me of some of the "hand-built" communication rigs I have seen radio amateur clubs throw together for their annual field day. It seems to be a collection of all of the usable communication equipment that could be scrounged and put together to serve some useful function. Because of my Minuteman work, I found myself mentally contrasting the City's emergency van with, for example, the Minuteman Transporter-Erector, also a special mobile unit; but, one that was designed from the ground up via the classic system analysis, preliminary design, etc. The contrast is profoundly indicative of what the nation's taxpayers are willing to spend for military equipment, but, reluctant to spend for city government's needs. During our tour of the communication facilities I was repeatedly impressed with the need for a quantitative evaluation of the City's present communication needs. Until that kind of measurement is done, and the results conveyed to the public in convincing terms, it may be that attempts to achieve increased appropriations for using new technology will fall on deaf ears.

. . . as indeed they do in most cases.

This engineer's comment, I believe, summarizes the City's problem. Not only in radio communications, but in sanitation, traffic control, water resource management, zoning, building codes, tax systems, air pollution, mass transportation, and many of the other problems facing our beleaguered city managements. Without the best talent in the nation addressed to the analysis of requirements, specification of system designs, and justification of their acquisition on a basis of maximum value delivered for every dollar spent, proposals for improvement will continue to fall on deaf ears. Now, it is a fact that most of the nation's research personnel are trained for and employed by other users. Furthermore, it is not fair nor efficient to impose research and development responsibilities on operational personnel. When the federal defense agencies realized this, a wide variety of research and development organizations were created and funded to provide R & D support for the military. The result is the most potent military capability in the world. If rapid progress is to be achieved in rejuvenating our cities, the same kind of massive R & D effort will have to be applied to their problems and a host of appropriate organizations utilized. Existing organizations will have to be redirected and new organizations will have to be created to insure the success of action programs.

However, it is not going to be easy to redirect existing talent or to create new talent to meet the City's need. There are several reasons why this is true. A first and important reason is money; or, more appropriately, the absence of money. DOD and NASA spend about $20 billion a year on research development, test, and evaluation. Last year the Department of Housing and Urban Development had an R & D budget of $10 million, much of which went to fund the new Institute for Urban Development. Furthermore, military and space R & D money is spent in a performance-first environment. To the engineer or scientist this means that he can practice his profession in an almost ideal environment; this constitutes intense competition for cities trying to attract skilled people. Not only are massive funds available to design and build the best high performance weapon systems, but the educational background of scientists is attuned to the objectives. DOD, NASA, and the industrial community that serves these agencies ask him to work on projects involving just these things.

What a contrast this presents when compared to the City's environment! At present, the City's problems seem to the scientist to offer nothing that is equally stimulating and intellectually rewarding. Few in local government or our academic institutions have been able to adequately interest him. Furthermore, little money is available at the local government level to attract him—one of the Technical Services Corporation's contractors said that the business they derive by serving local government is so small that they are embarrassed to talk about it—and, there are few places for research scientists to work where their efforts will have meaningful and direct impact on the problems of the city.

We do, however, have some things working for us. Many competent engineers and scientists have a social consciousness, a sense of responsibility to the society that provided their education; if we can provide a reasonable working environment, appropriately funded, top-level people should be intrigued by the challenge inherent in applying their knowledge of science and technology to the problems of the City.

Our experience to date with the Technical Services Corporation seems to support this hope. In February of this year, Mayor Yorty asked the Technical Services Corporation to prepare two Model Cities grant requests for the City. The deadline was two months away and the Corporation had no appropriate staff. Volunteers were recruited from five local corporations: economists, psychologists, system integration specialists, scientists, and planners. Various government agencies provided partial funding for the effort, the Corporation committed its own funds, other specialists were retained, and community organizations offered cooperation. With six weeks to go until the deadline, we put a dozen professionals in a room and told them to kick a field goal from the fifty yard line. The result was inspiring to watch. The collection of people from industry, City government, and the community, and the formulation of proposals, was similar to the normal proposal preparation process that occurs in most large aerospace corporations. However, this was the first time an activity like this took place in city government, on a social program. The proposals were written and released on time, after six weeks of concentrated research. They received the unanimous support of the City Council, and the Los Angeles Chamber of Commerce, and the endorsement of the Mayor. Our point was made. Scientists can and will, under appropriate circumstances, concentrate their efforts on city problems.

In its short existence, the Corporation has also been able to initiate a project to develop communication systems for emergency service agencies and to assist the City with design and management of its Community Analysis Program. We have found the talent of industrial scientists and engineers immediately applicable to these tasks, and the individuals eager to help with the problems of local government.

In closing, it is worthwhile to describe the Corporation and its objectives. It may be one of the organizational vehicles necessary to bring substantial progress in the urban environment.

The overall objectives of the Corporation are to establish and maintain a highly competent, technically oriented organization at the local government level. Using the potentials of modern technology, it provides a research and development capability for local government—a capability that has been nonexistent.

The Corporation provides training in modern techniques of system analysis for city personnel. Development of this expertise will allow better communication between local government and the industrial community. The Corporation serves the function of providing the interface between government and industry to assist in the development of an understanding of the complexity of programs involved.

The Corporation does not compete with industry, but rather assumes a program manager's role similar to that of the aerospace corporation on developmental programs, and provides industry with guidance so that efforts are channeled to problem areas.

Since our experience indicates we are on the right track, we are expanding the base of support for the Corporation. We are attempting to apply these techniques to other urban problem areas, such as:

1. Utilization of the Highway Development to reinforce economic development of cities.
2. Compound attacks on problems—the analysis of the need for programs of a social and economic nature to complement physical or housing programs.
3. The development of techniques to better utilize the information available to cities.

The Corporation has received financial support from Los

Angeles and from philanthropic agencies, and we are working to expand this support. We expect systems developed by the City and the Corporation to be applicable nation-wide, and we look forward to the time when opportunities will exist to use our results for the benefit of other cities.

The "Evaluative Function" in Government*

NICHOLAS E. GOLOVIN

Executive Office of the President
Office of Science and Technology
Washington, D.C.

In searches for rational policy alternatives concerning even clearly circumscribed sectors of governmental activity, one sometimes finds it difficult to reach satisfactory conclusions without expanding the "universe of discourse." And, the broader the sector and/or the higher the level of government, the more likely is it that useful results will not be obtained without viewing at least some of the aspects of the problem of interest from a national point of view. The concept of an evaluative function and, specifically, an evaluative branch of government arose in this way during my involvement with several apparently well-delimited problems in the management of federal research and development programs. In exploring the implications of the idea it became apparent that the concept promised improvement in capabilities for coping, at least theoretically, with many important socioeconomic problems currently concerning the several levels of government in the United States.

* Portions of this paper appeared in *Management Science: Application,* June, 1969.

157

The discussion begins with a brief description of some social problems whose systematic analysis and treatment today are inhibited by the absence of suitable fact-gathering and analytical machinery. It then proceeds to a discussion of the several reasons why such machinery is not now available, and of some of the alternative ways in which it might be feasible to provide it. The concept of an evaluative branch of government is then explained and some of its potential advantages outlined.

As should be clear from the nature of the subject, the views expressed or implied are those of the writer alone.

Growth in population, advance in technology, and development of techniques for effective organization and direction of large aggregates of people and resources appear to have transformed the national economy of the United States, and those of several other countries, into complex, internally highly integrated *systems*. As *systems* such national economics have the characteristic that events in relatively narrow sectors produce effects which propagate broadly and cause significant, sometimes unexpected, disturbances in widely separated areas.* In addition, because technology now provides the basis for many economically important activities, and because its evolution, while not predictable, seems increasingly rapid, the existing, relatively static, institutional frameworks tend everywhere to be subject to more intense stress. Examples of such difficulties are: (a) growing recognition that many acute social problems which, while seemingly of local origin (e.g., urban congestion, poverty, civil instability), are strongly coupled to

*Editor's note: This is an important insight. In my opinion, this property seems to be present in most large-scale systems. It can imply (by a chain of reasoning not given here) that certain types of instability can exist which may easily elude detection (since they may be a property of the system *as a whole* —and not of any individual component). Such instabilities can suddenly make themselves known in the form of an unpredicted catastrophy. Examples are: power system breakdowns (see S. Brown's paper in this volume), telephone system breakdowns (N. Wiener, *Cybernetics* 2nd edition, MIT Press, 1961), p. 150-151, economic system breakdowns (the depression), etc.

One recurrent theme of this meeting was: to manage such systems, one must use a method of control which involves a *coordination* between the subsystems. In the present case (and this was also pointed out by Savas) interagency coordination is far too low at the present time in governmental systems.

various internal characteristics of the socioeconomic system as a whole; (b) the need to devise, under limited overall budgets, rational and generally persuasive bases for setting relative priorities among national programs (*viz.*, urban redevelopment, new systems for urban and regional transportation, major improvements in the educational level of large segments of the population); and, in particular, (c) devising a broadly acceptable basis for deciding how the costs of such programs are to be allocated among the income-producing elements of society.

One of the distinguishing characteristics of such problem areas is the difficulty of reaching understanding and doing something about them through existing organizational arrangements at any level of government. This is so, at least in part, because the scale and scope of such problems require inter-agency analyses and planning while agency responsibilities are circumscribed and resources inadequate at the municipal, state, and even federal levels. Furthermore, generally satisfactory solutions will inevitably involve the active participation of non-governmental organizations whose individual roles and associated motivational arrangements will need to be specifically defined and provided for—tasks for which there are neither resources nor a generally accepted guiding philosophy.

The fact that the anticipation and resolution of major socio-economic difficulties require that *society's activities must be viewed in their totality* is reflected in the increasing frequency of pleas in Congress and elsewhere that the so-called systems engineering skills developed in the aerospace industry be applied to deepen and broaden existing analytical capabilities at the municipal and higher levels of government. For example, a bill was introduced in the 90th Congress to establish "An Office of Program Analyses and Evaluation and a Joint Committee of Congress on Program Analyses and Evaluation." [1] This bill proposed that the Office "shall make a full and complete study and evaluation of all Federal Programs and activities," and that the Joint Committee "shall make continuous studies . . . of the problems relating to the effectiveness of Federal Programs and activities and the establishment of priorities for public funds." Similarly, the National Commission on Technology, Automation and Economic Progress, in its review of some of our main social problems at the national level, expressed the conviction that "a new intellectual technique, that of system analysis, can provide a new approach to meet government planning needs." [2] This finding is, in part, the result of the Commission's examination of the questions "How can human and community needs be met?" and "How can they be more readily recognized and agreed upon?" and its conclusion:

> What concerns us is that we have . . . no ready means for agreement, that such decisions are often made piecemeal with no relation to each other, that vested interests are often able to obtain unjust shares, and that few mechanisms are available which allow us to see the range of alternatives and thus enable us to choose with a comprehension of the consequences of our choices.[2]

As the final example, Vice-President Humphrey, in a speech at the Seventh Annual Goddard Dinner of the National Space Club, challenged the aerospace community to help alleviate urgent social problems by applying its special skills in systems analysis and operations research. As a result, the Operations Research Society of America and the American Institute of Aeronautics and Astronautics held a joint Forum on Systems Analysis and Social Change in March 1968. The central conclusion of this Forum is said to have been the growing importance of the need:

> To invent and to introduce into our existing arrangements for national management (constitutional democracy plus the free-enterprise system) a subsystem which will evaluate how well the total system works, and which will then generate "error signals" for activating pertinent corrective reactions throughout the over-all management process.[3]

The increasingly important role of technological change as one of the main wellsprings of economic and social change has, of course, been recognized for a long time. The introduction of computers into the business system on a large scale in the 1950's was associated with the emergence of intense public interest in the subject of Technology and Social Change. Recent years have been marked by steadily increasing efforts

in government and the universities to achieve deeper understanding of the subtle and exceedingly complex interactions
among progress in science, advances in technology, the rate of
technological innovation in industry, and the technical and
economic consequences of such innovation. One of the major
efforts along these lines was begun in 1964 by Harvard University in the Program on Technology and Society supported
by a substantial grant for a ten-year period by IBM.[4] While
this program has some six years to go, E. G. Mesthene, its
executive director, has said recently:

> Among the effects of technological change that we are beginning to
> understand fairly well even now are those (i) on our principal
> institutions: industry, government, universities; (ii) on our produc-
> tion processes and occupational patterns; and (iii) on our social and
> individual environment: our values, educational requirements, group
> affiliations, physical locations, and personal identities. All of these
> are in movement.

And, he concludes his discussion of "how technology will
shape the future," by pointing out that:

> The most fundamental *political* task of a Technological world . . .
> is that of systematizing and institutionalizing the social expectation
> of the changes that technology will continue to bring about. . . . We
> need above all, in other words, to gauge the effects of technology on
> *polity*, so that we can derive some social value from our knowledge.[5]

As was noted above, the National Commission on Technology, Automation and Economic Progress recommended
systems analysis as a "new approach to meet *government planning needs*" (italics added). Since the essential characteristic of the systems point of view is extension of analysis
and design to the entire complex of interacting elements and
constraints involved in the successful operation of a complex
aggregate, the linkage of systems analysis to government planning has important implications if we also recall the pertinent
meaning of the term plan: "method or scheme of action, procedure or arrangement." [6] Since the main social problem areas
are deeply interrelated and involve practically the entire economy, the government is basically being urged by the National

Commission to use systems analysis in developing a scheme of action *for the national economy as a whole.*

An interesting question arises immediately: is, and if so, to what extent is the government, the federal government specifically, now engaged in systems analysis particularly as applied to planning?

Of course, the answer is *yes* and in most agencies. After some four years of development and application in the Department of Defense, the Planning, Programming and Budgeting System (PPBS) was introduced into the civilian agencies of the Executive Branch in 1965. An authoritative discussion of the nature, advantages, and weaknesses of this system was provided earlier this year by Charles Schultze, a former Director of the Bureau of the Budget, who had much to do with the development and introduction of the PPBS for use by the civilian agencies. The 1968 Rowan Gaither Lectures were delivered by Schultze and were concerned mainly with the PPBS.[7]

Fundamentally, the purposes of the PPBS are these: (1) make the goals and objectives of each important area of governmental activity an explicit matter of concern and analysis; (2) enforce agency analysis of past and future programs so that their outputs can be related to specific program objectives; (3) require definitions of objectives and formulations of programs for a time span substantially longer than the single year to which annual budget submissions apply; (4) insure the development of useful estimates of total program costs—not just for the budget year when they are proposed to begin; (5) require the definition and analysis of alternative plans for achievement of program objectives as a process for helping establish the optimum course; (6) establish this set of procedures as an essential, systematic part of the annual budgeting preparation and review process.

In presenting his estimates of the current and future value of the PPBS, Schultze begins by setting forth in considerable detail the main objections by Lindblom [8] to the substance as well as to the underlying philosophy of the system. For the

purpose at hand the main lines of Lindblom's attack, as formulated by Schultze, can be summarized as follows.

(1) PPBS' emphasis on making goals and objectives explicit
is counter-productive because the first law of successful politics
(certainly in the U.S.) is that debate about objectives, ends,
and means should be minimized; and, in any case, the realities
of the decision process are not compatible with the problem
solving orientation of the system; (2) Not only is the PPB System incompatible with the realities of the political process, it
does not specify the *optimum* process for good decisions in a
free political society; in particular, its attempt to introduce the
quantitative approach into specification and measurement of
means and ends, and to force the separation of means from
ends through definition and analysis of wide ranges of alternatives, is neither desirable nor attainable in governmental confrontations with social and institutional problems. Schultze's
measured defense of the PPB experiment is specific and over-
whelming. It is basically this: (1) Participants in the decision
process have to be informed about the relationships between
program inputs and outputs, and the connections between
outputs and social objectives—otherwise the political bargaining process will simply mill in a welter of ignorance; (2) Since
the role of the federal government has in recent years expanded
into areas in which relationships between program inputs and
outputs have become increasingly complex, it is nonsense to
expect that rules of thumb, political intuition or even past
experience can provide adequate bases for responsible governmental decisions; (3) Detailed comparisons of alternatives, analytically related to evaluations of past programs, rather than
interfering with the adversary process in political debate will
support and strengthen it.

How would Schultze improve the PPB System and, more
generally, the associated governmental process? The main suggestions are: (1) implement the multi-year aspect of the system
by inviting agency submissions of plans for programs requiring
budgeting decisions beyond the budget year under discussion
(currently agencies' future PPBS plans include only the consequences of decisions in the budget year at issue); (2) enforce

the definition of agency programs so that they take full account
of the existing private and governmental incentive structures
in the sector of society involved; and include the additional
positive incentive mechanisms required to effectuate program
objectives; (3) introduce an overall regional structure for the
federal role at the state and local levels providing a regional
presence for the executive office of the president, and capable
of dealing with problems arising from the operations of co-equal
executive agencies in the same geographic areas.

Noting the increasing importance of inter-agency program
planning and implementation issues emphasized in our initial
discussion, and clearly recognized by Schultze, it is necessary
to point out an additional and perhaps crucial shortcoming of
the present version of the PPB System. It provides no locus
or comparable procedures for the definition, analysis, and evaluation of alternative program plans for the government as a
whole. That is, the system does not squarely confront the
program priority problem *above* the individual agency level.
And this is the level which appears to be most involved in
defining the character of national responses to problems which
are not bounded by particular technical disciplines, agency
responsibilities, and geographical or political jurisdictions.

The program priority is especially relevant to another proposal for improvement of the government's analytical and
planning capabilities, which is useful for us to consider. This
proposal is developed in a report by Gerhard Colm and Luther
Gulick prepared for the National Planning Association.[9]

To begin with, these authors identify a number of difficulties
associated with present methods of federal involvement in
various social problem areas. For example: (1) the official list
of federal assistance programs to local governments is 701 pages
long and contains descriptions of 459 programs so that "a major
research effort is needed if a city manager wants to find out if
there is a federal or state grant-in-aid program from which he
can obtain financial support to deal with a specific social problem; (2) on the other hand, it is not clear to what extent, if at
all, federal plans and policies in various problem areas take
account of related activities and plans of state and local govern-

ments or those of the private sector; (3) neither does the federal government appear to take account in its planning of the "impact of government programs on private enterprise" and because "in a pluralistic society . . . both the public and private sectors contribute to the definition and pursuit of national goals . . . government programs must . . . take into account both the contribution private enterprises and other private institutions can make toward the achievement of a desired goal, and the effect (positive or negative) that a government program will have on activities in the private sector." It is especially interesting that the last two points say almost exactly the same thing that Schultze emphasized in his proposals to improve the PPBS—federal programs do not now, but must eventually, take into full account the pertinent motivational structures in areas where they have significant impacts.

Secondly, as concern with difficulties noted above suggests, these authors believe that what is needed is *not governmental* but *concerted* planning. Such activity would be an amalgam of: "(a) planning of governmental policies and programs, (b) planning of investment and marketing programs by individual private enterprises, and (c) planning for social and other improvements by numerous private groups and organizations." Thus, such joint efforts by government and the private sector would appear analogous to what has been called consultative planning, a technique reputed to have been developed to a relatively high level of usefulness in Western Europe, especially in France.[10]

Finally, Colm and Gulick believe that self-consistent evaluation of the direct and indirect costs of various competing programs in terms of potential claims on the GNP and on manpower will be essential if priority analysis is to be meaningful in other than purely subjective terms. And to attain such self-consistence it may be necessary to develop new systems of social indicators useful, on the one hand, for helping appraise the social utility of particular programs and, on the other, for building bridges between their apparently incommensurable outputs.

This analysis leads Colm and Gulick to two principal recommendations: (1) To provide a link between the President, his administration, and the pertinent groups in the private sector, a "Citizen's Committee on National Goals and Priorities" should be established; (2) there should be established "the post of a new Special Assistant (or Special Counselor) to the President for Plans and Priorities." He would direct an "Office for Appraisal of National Goals and Programs," and provide the means for "the evaluation of Federal programs in the light of the *President's vision for the nation's future* (italics added). This office would be an independent entity in the executive office of the president and would be established by statute specifically for dealing with long-term goals and programs as a safeguard against its absorption into short-term assignments.

One of the concluding paragraphs of the Colm/Gulick report concerns the timeliness of these recommendations. It says:

> Thus, we conclude that the time is ripe—in terms of both need and capacity—for a further step toward concerted planning. The problems created by the worldwide responsibilities of the U.S., by the economic bottlenecks and social tensions resulting from rapid technological advances and slow adjustments in institutions at various levels of government, by the desultory adoption of a multitude of programs, each in response to a specific need and special pressure—these developments impose the stern necessity to make priority decisions of grave consequence. This country can no longer afford the luxury of deciding on programs without full recognition of their respective urgency in terms of national goals.[9]

A pertinent but obvious question, provoked by the preceding discussion, is: why have no means been so far provided for government agencies, at any level, for dealing effectively with long-run problems which do cut across existing agency and institutional lines?

A large part of the answer is that, for the government, planning is and apparently has always pretty much been a dirty word. Today, for example, the business community accepts all concepts of goal setting, systems analysis, planning, and control for organizations in the private sector of the economy but not for any level of government. It seems to be a matter of basic ideology that a systems approach to an enterprise implies greater efficiency and profits; while the same approach by government to any important phase of national life necessarily means inefficiency, a multiplicity of rigorous controls and unnecessary losses of freedom in all kinds of individual and corporate decision-making. And the main obstacle to inter-agency programming and planning at the federal level has probably been the inflexibility inherent in the Congressional committee structure and the special relationships prevailing between agencies and their committees.

Since it seems clear that ways will need to be found either to reeducate the great majority of Americans that such views are incorrect and must be replaced by love for federal, state, and local planning (which is probably absurd), or to devise popularly acceptable techniques for attaining many of the potential benefits of governmental planning without actually doing it (which may be feasible); it is useful to dwell on the underlying issue in some detail. A convenient way to begin is to appreciate recent business community reactions to the thesis argued by Galbraith in *The New Individual State*.[11] We'll consider two of these.

Galbraith's main thesis can be summarized as follows: Although the central purpose of the U.S. economy is growth in revenue (economic goods) at the national and personal level, this purpose is self-defeating. This is so because continuous economic growth implies a rising educational level for the people who, as they became more sophisticated, increasingly think less of material rewards and more of "questions that are beyond the reach of economics—the beauty, dignity, pleasure, and durability of life." And, therefore, it is a large expansion in the so-called public sector of the economy rather than growth in the output of business enterprise that is wanted and needed. Furthermore, because the national economy is already planned by the large corporations in cooperation with the central government (which limits its own role to using fiscal policy for stimulating economic growth, educating the labor force, and supporting research and development), what is needed to set things on the right track is a change in the character of the prevailing national goals and in assignment of responsibility for planning.

Fortune attacks Galbraith's conclusion on two counts. It says, first, the economic reasoning is unsound, and, second, a planned economy involves abandonment of the free markets concept and necessarily implies totalitarianism and the disappearance of free enterprise. The heart of its argument goes as follows:

> Economic theory, as we conceive it, stands or falls with the idea of consumer sovereignty. . . . Enterprise involves uncertainty about how the consumer will behave in the market place; if the consumer's behavior is going to be 'managed,' then entrepreneurial risk-taking becomes a contradiction in terms. . . . Thus a 'planned economy' necessarily implies the 'abandonment' of free economic enterprise. . . . For if you have a government that profoundly believes it knows what people should want, regardless of what people actually do want, then fine calculations of marginal utility are quite beside the point and there is no place for economic enterprise of any kind.[13]

Business Week's argument is essentially identical except for one striking difference. Although *Fortune* assigns a sovereign role to the consumer and, by so doing, implies that the totality of consumer's wants can be equated to the prime goal of the U.S. economy, *Business Week* is not so sure that this is the case. The review says: "Yet, even if the academics chop away part of Galbraith's vision of modern industrial life, it is doubtful that they can diminish the popular appeal of the Galbraith cry for more attention to social values. *The perplexing question, though, is how to direct—or redirect—economic activity so that it better serves the goals of the American people"* [13] (italics added).

Thus, it is implied that there are goals for Americans which are not consumer/market-determined, and that it might be important to shape economic activity so that they are attained. However, it is also implied that perplexity arises as to what should be done because maintenance of the free market system is (or should be) one of America's goals.

Thus, to the extent that the views of these *Fortune* and *Business Week* reviewers can be assumed representative, there seems little question that the business community's desire for economic *autonomy*, i.e., for freedom to invest, buy, produce, price, employ as its own self-interest deems best, remains unshaken by any of the technological, social, or economic developments of recent years which have been alluded to above.

Furthermore, as a review of Lindblom's arguments will show, it is not necessarily only naked self-interest which is responsible for such business community or other groups' views. For example, it seems to be true that social ends and means are sometimes difficult to separate and, as Schultze explains in the reference previously cited, useful separation and weighing of relative values can frequently be accomplished best not in advance as planning requires but through the actual process of resolving a problem situation. Also, in part because of the absence of sufficiently complete data, and in part because of the extreme complexity of social dynamics, there are serious difficulties in predicting the consequences of program inputs as to the nature, magnitude, and values of resulting outputs. Thus, the problems of prediction in social systems analyses can be incomparably more difficult than in the most complex of hardware systems. And, it follows therefore, that the slow, incremental approach to social progress realized through pluralistic political bargaining may be no less efficient for such purposes as the market has been for determining prices and allocating resources.

Now, the important point about this line of argument for us is not its validity which, evidently, can neither be proved nor disproved, but its general plausibility and the degree to which it might be accepted within the politically effective constituencies. However, these are matters of largely subjective judgment, and the discussion above already implies a spread of widely different views even among the experts. Thus, convergence to a definite conclusion necessary for further progress is not possible without additional insights. With this in mind, let us look at the institutions of the presidency and the Congress, each of which will clearly be affected by proposed changes in governmental organization at all levels. And to sharpen the pertinent issues, let us focus on the Colm/Gulick proposals. That is, let us examine the effects on, and the interests of, the President and the Congress in the adoption of their proposals.

One of the fundamental characteristics of the American

governmental system is the presence of checks and balances, whose function is to limit the growth in power of any major institution at the expense of others. Such limitation arises from conflicts engendered by the sharing of powers by separate (and independent) institutions of which the presidency and the Congress are the most conspicuous examples. In general, a balance of power is maintained between such institutions over the long run although one or another may attain dominance for more or less prolonged intervals. For example, although since the time of the second Roosevelt the presidency appears to have acquired power for greater initiatives in domestic affairs (e.g. through an expanded and strengthened executive office of the president—Bureau of the Budget, Council of Economic Advisers, Office of Science and Technology), the Congress has over the same period increased its control over executive agencies through the requirement for annual authorizing legislation for many agencies, and by delaying the enactment of appropriations until well into the appropriation year.

According to Neustadt: "Effective influence for the man in the White House stems from three related sources: (1) the bargaining advantages inherent in his job with which to persuade other men that what he wants of them is what their own responsibilities require them to do; (2) the expectations of those other men regarding his ability and will to use the various advantages they think he has; (3) those men's estimates of how his public views him and of how their publics may view them if they do what he wants." [14]

Now let's assume that this is a realistic description of the basic elements in presidential effectiveness, and that there were added to the executive office an adequately staffed "Office for the Appraisal of National Goals and Programs" and a "Citizens Committee on National Goals and Priorities" to provide a link between the president and power centers in the private sector, as Colm and Gulick propose. Then, almost on the face of it, two conclusions would appear justified: (1) the president would (and should) deliberately use these new resources in various ways not only for the performance of their substantive functions but, and perhaps more importantly, for increasing the effectiveness of his leadership, as well as for magnifying his bargaining influence relative to all others having a voice in the national management system; (2) in the total structure of "separate institutions sharing powers" (Neustadt's description), the presidency's share of powers would be significantly increased while all other shares, and in particular those of the Congress and the business community, would be correspondingly decreased. And it then follows that, on the basis of expectations by the Congress and the business community that these two outcomes with respect to changes in presidential powers would be inevitable, they would use all of *their* influence to insure that such changes did not occur in the first place.

Furthermore, this structure of separate institutions sharing powers has built-in symmetry properties which equally well support the conclusion that, if it were proposed to invest the Congress rather than the presidency with the planning and appraisal functions under discussion, then the business community would join the presidency in exercising their joint influence to prevent the anticipated change in the existing balance of shared powers.

Finally, we can conclude this discussion of the difficulties facing proposals for the introduction of planning and evaluation functions into the existing governmental structure by referring again to the report of the National Commission on Technology, Automation and Economic Progress:

> Our concern is to strengthen this system (i.e., the U.S. system of free enterprise and constitutional government) at a time *when social and technological changes begin to confront us so directly* and when *we need some means* of *assessing the consequences of such changes in a comprehensive way.*
>
> The future is not a task for government alone. In fact, *the concentration of forecasting mechanisms entirely in the hands of government, particularly at a time when such forecasting becomes a necessary condition of public policy, risks one-sided judgments—and even suppression of forecasts for political ends.*
>
> Along with forecasting there *is a need to set national goals and to enlarge the participation of all sectors in the public debate which would be necessary in the statement of priorities.* For this reason the Commission, while not endorsing any specific format, feels that *some*

national body of distinguished private citizens representing diverse
interests and constituencies and devoted to a continuing discussion of
national goals would be valuable. Such a body would be concerned
with *"monitoring" social change, forecasting possible social trends,*
and *suggesting policy alternatives to deal with them.* Its role would
be *not to plan the future but to point out what alternatives are
advisable and at what costs"* [2] (parenthetical note and italics added).

In the above statement, the Commission appears to be using
the terms government and private citizens as mutually exclusive
categories, with government being a synonym for the combination of the existing executive, legislative and judicial branches
at the federal level, the point being that the essential characteristic of private citizens in the Commission's statement is
that they take no part in running the existing government, a
distinction to which we will come back in a later section.

Focusing attention, first, on the question of what are the
special capabilities the overall national management system
requires in order to react effectively to technological and
social change, the various recommendations set forth and
implied in the discussion of the preceding sections can be
summarized as follows:

(a) American society urgently requires the invention of a
new organizational mechanism, preferably one which could
be applicable at all levels of government (i.e., local, state
and federal), to: (i) collect, interpret and analyze information; (ii) define potential problems and needs; (iii) develop
pertinent alternative action plans, including detailed
analyses of the associated cost/benefit effects, for consideration by the responsible branches of government, the private
business sector and the public; (iv) systematically, and in
real-time, evaluate the results of established programs in
comparison with the goals that had been envisioned for
them at their inception and develop recommendations for
changes in program content or implementation; and (v)
keep the government, the business community, and the public as a whole informed in depth of the nature and results
of its work.

On the other hand, focusing attention on the realities of

how the structure of separate institutions sharing powers functions in national system management, we can summarize the pertinent arguments of the preceding sections thus:

(b) A mechanism possessing the magnitude of resources and the continuity of life probably required by the scope, persistence, and controversial nature of the issues with which it would be concerned—at any level of government—would not be accepted by the people generally and by the business community particularly as part of the existing branches of government (specifically, of the legislative and executive branches).

And the central question now becomes how are the needed additional capabilities to be provided so that there is minimum disturbance to the existing balance of powers in the institutional structure. On the face of it, the new mechanism must therefore be: (1) outside the existing governmental structure; and (2) relatively powerless *vis à vis* the main existing institutions inside and outside the government. What, then, are the reasonable possibilities? They are: a private profit-making institution; a university; a not-for-profit institution; a commission of distinguished citizens; a new, i.e., a fourth, branch of government.

Private profit-making corporations doubtless could be quite effectively used for performing several of the functions itemized in (a) above, and perhaps even all of them at the local level of government for relatively small and/or isolated communities. However, it would seem difficult to advocate the public acceptability of such a mechanism at the large city, state, or federal levels. The recent entry of large corporations, particularly from the aerospace industry, into contracts at many levels of government for the application of systems engineering capabilities to social problems has been to provide specialized technical or managerial services rather than to engage substantively in over-all planning or in the evaluation of community goals and of programs for their attainment.

The university has recently become the focus of attention as the institutional form for helping society out of its growing

difficulties. For example, Walter Lippmann is said to have expressed the view that the university is "the only institution left
to fill the void left by dissolution of traditional authority in the
modern, secular world." [16] In a similar vein, while reflecting
on the increasing uneasiness of people in their interaction with
the "growing complexity and impersonalization of . . . society,"
Glenn Seaborg has said: "To me these feelings forecast the
need for a huge reevaluation of our goals and values, and it
will be in our universities where such reevaluation will take
place." [17] In the sense that the long-run cultural development of
society is largely the product of research, scholarship, teaching,
and other intellectual interactions of university communities,
and of social leaders trained in them, with society as a whole,
such views can hardly be questioned. However, a number of
serious objections can be raised as to the suitability of the university as the organizational device for the functions outlined in
(a). In the first place, the fundamental and irreplaceable role
of universities in cultural growth and evolution can probably
be maintained only if their basic institutional character (as
centers for research, scholarship and teaching) is not seriously
compromised by massive involvement in activities of other
kinds. Thus, while the Carnegie Corporation for the Advancement of Teaching in its 62nd annual report urges the "bringing of an academic institution's special competence to bear on
the solution of society's problems," it also warns that doing so
has many risks and that "public service activities could, if uncontrolled, dwarf the university's other pursuits," [18] thereby
alienating both students and faculty. Secondly, the *direct* involvement of a university in complex undertakings (in contrast
to relatively indirect roles in such work as a funding channel,
or as a legal shelter from direct governmental involvement in
project staffing and management—as in the case of some large
weapons or space systems laboratories) is questionable in part
because of inappropriateness of university organizational forms
(i.e., structuring by discipline) and emphasis on specialized
rather than interdisciplinary competence as the preferred route
for professional achievement. This line of argument concerning university involvement in large technological projects and

its extension to major future undertakings in the life services
has recently been strongly supported by Harvey Brooks:

> The development of complex systems involves the coordination of
> many component pieces of a problem and many individual speciali-
> ties. Often it involves highly sophisticated science or mathematics
> side by side with rather conventional or mundane design or repetitive
> analysis. Such a coordinated effort tends to be incompatible with the
> university environment, with its high turnover of people, with its
> treatment of research as a part-time activity, and with the high value
> it places on individual as opposed to team performance, and on the
> proposing of new ideas as compared with critical evaluation and com-
> parison of ideas and their execution in all the most mundane detail.
> *In the future, we may expect more enterprises in the life sciences to
> partake of the same complexity that is now characteristic of many
> engineering systems. Thus the increasing significance of "intellectual
> size" in these areas may generate greater reliance on mission-oriented
> institutions only loosely associated with universities or completely
> separate"* (italics added). [19]

The not-for-profit corporation, as an organizational device for
bringing together high-quality intelligence and training for
interdisciplinary confrontations with complex technological,
economic or sociological problems, evidently is the most suitable device available. And there is probably little question that
no matter how our society ultimately provides a suitable
mechanism for the functions outlined in (a) above for the
national and state levels of government, the not-for-profit think
tank will be extensively used as the main such mechanism at
the local level and as an important source of technical support
at the national and state levels. The principal reason for questioning its suitability for the higher levels of government is
simply that as a permanent institution serving the needs, for
example, of a very large municipality, a state or the federal
government, it would probably meet insuperable difficulties in
maintaining intellectual freedom and relative political independence over the long run. In the first place, for the large
centers of government, demands for the think tank's resources
would be so great that holding on to the contract would necessarily become a major year-to-year preoccupation of both management and staff. In the second place, as does not seem to be
widely appreciated, systems analysis is not intrinsically an

ideologically or politically neutral technical discipline when
applied to most governmental problems. It approximates these
characteristics when the system of interest is a combat-ready
division, a guided missile or a moon-ship, but cannot do so even
remotely when the system is a large socioeconomic aggregate.
Thus, although systems analysis is generally considered a later,
and therefore more advanced, stage in the development of
techniques called operations research, it usually tends to be
much more qualitative in character. As has been recently summarized by Wildavsky:

> The less that is known about objectives, the more they conflict, the
> larger the number of elements to be considered, the more uncertain
> the environment, the more likely it is that the work will be called a
> systems analysis. There is more judgment and intuition in systems
> analysis, and less reliance on quantitative methods, than in operations
> research.[20]

And the point of interest to us is that the basically qualitative
character of the work, combined with inevitable pressures from
the contracting authority not to produce facts or recommendations which directly or potentially weaken its public program
posture or its bargaining position in the power structure, could
result in greater stresses on the think-tank than might be acceptable from the viewpoint of general public interest. Of course,
this argument is much weaker if the think-tank's involvement
with any particular contracting officer calls for a relatively
small fraction of its total resources. But in such cases—unless
not-for-profit think-tanks get to be much larger organizations
than appears to date to have been found feasible to establish
and manage well—its services would be to a relatively small
community or be of a largely supportive character in case of a
larger one.

The use of commissions of distinguished citizens for coping
with various kinds of governmental problems is an old and
well-established device. Commissions have been used for many
purposes (advisory, evaluative, fact-finding, public relations,
policy recommendation, etc.) and for several different kinds of
functions (e.g., to permit representation in government by special constituencies, to establish the maximum policy tolerance

levels of special interest groups in key issues, and to provide elite participation in the formulation of public policy) . There is little question that this device has in the past been useful at all levels of government, and that it is currently the only available, publicly acceptable technique for achieving and using a kind of consensus on major socioeconomic problems at the national level. As Daniel Bell has put it, "government by commission" can today make an important contribution because:

> The government today does not have any single agency which seeks to 'forecast' social and technological change (though an individual agency, such as the Census Bureau, may be concerned with population trends, and the Council of Economic Advisers with short-run economic projections). There is no agency which seeks to link up current and possible future changes in a comprehensive way so as to trace out the linked effects on different aspects of government policy. *And perhaps most importantly, at a time when we must begin consciously to choose among 'alternative futures,' to establish priorities about what has to be done—for it is only an illusion that we are affluent enough to take care of all our economic problems at once—we have no 'forum' which seems to articulate different national goals and to clarify the implications and consequences of different choices.* The Congress is not such a forum. . . . 'Congress does not resolve national controversies; it can only act after most of the controversy has been resolved' . . . (italics added).[21]

Although Bell's list of the functions for which no government agency is now responsible (at any level of government, although his argument is aimed exclusively at the national level) is essentially identical with those itemized in (a) , his conclusion that the commission mechanism is useful in these connections cannot be interpreted as a *recommendation* that such functions should be handled by commissions. Although he does not say so, his argument seems clearly to imply recognition of (b) — namely that *the non-existent agency with the needed functions cannot be created as part of one of the current branches of government; in the absence of a more suitable mechanism, he merely concludes that commissions can play a useful role.* But there are several important reasons why it is unreasonable to expect that government by commission will meet the most important needs among those in (a) above or as listed by Bell.

In the first place, commissions are government bodies and, if assigned responsibilities of interest in this context, would

usually be part of the executive branch at the level of government using them. Therefore, if (b) is not to be violated, they
must always be temporary. But the main problems of interest
are clearly of continuing character, and it would seem, on the
face of it, unreasonable that a new commission is to be created
from time to time to re-review problems in areas of vital and
unending importance. Second, unless we conceive of the establishment of super-commissions with larger membership, with
larger staffs, and with much longer lifetimes than has heretofore been found feasible and desirable, there are many problems which are probably not suitable for permanent "government by commission," at least at the national level—structural
poverty, civil instability, urban redevelopment, educational
opportunity and quality of education, the general problem of
establishing and rationalizing budgeting priorities among major
federal programs in all areas, are probably all of this kind.
Finally, there is always the non-negligible danger that an ap-
pointive body with a temporary life, particularly at the lower
levels of government, can be used by the existing power structure as a convenient device to obtain political support for
particular goals or for special techniques to ensure their
realization.

There remains now the alternative of modifying the institutional framework set by the existing three-branch structure of
government, and of creating a fourth branch to assume the
several governmental functions not envisioned or considered
important at the time the present constitutional framework was
originally devised.

What is proposed, then, is that the present three-branch structure of government (executive, legislative, and judicial) at the
local, state, and federal levels be supplemented by a new branch
entitled, perhaps, the evaluative branch, to be concerned
broadly with the functions suggested [in (a) p. 172] above.

To take account of the many kinds and sources of objections
to the assumption of such functions by one of the existing
branches of government, it would be necessary to help assure
the policy and functional independence of the new branch by

appropriate constitutional and procedural guarantees. It would also be essential to circumscribe its composition, its authority, operating procedures, and its interfaces with the other branches and with the public in various ways. Specific constraints would be necessary to assure the business community and the electorate generally that its principal objective would be to assist in the preservation and strengthening of constitutional democracy and of the free enterprise system by undertaking, under close and public supervision, those information, analysis, planning, and evaluative functions which it has not been hitherto feasible for other branches to assume, but whose performance has now come to be generally recognized as socially essential. It would also be important to insure initially that the new branch would not dilute any of the responsibilities of the existing branches, nor assume information, analysis, planning, or evaluative functions for which other branches are now responsible and which are necessary for their effective functioning. Thus, it would be made clear, as a phase of the process of establishing the new branch, that its activities would be *non-executive, non-legislative,* and *non-judicial,* and that *its concern would be primarily with the longer-range, inter-branch, inter-agency, inter-disciplinary issues which the existing organizational structure finds it difficult to face, to understand, and to resolve.*

This exclusion of the evaluative branch from the process of *direct* government, which has been, and would continue to be, the *sole prerogative* of the existing branches, effectively places the officers and staff of the new branch in the class of private citizens as defined, for example, by the National Commission on Technology, Automation and Economic Progress—provided, as previously suggested, private is interpreted to mean *operationally uninvolved in the processes of the executive, legislative, and judicial branches.** Therefore, if the body of distinguished private citizens which the Commission recommends is identified

---

*Editor's note: Of course there must be "checks and balances" in the overall structure in the sense that the other three branches of government must be in a position to challenge and "check up" on any information produced by this fourth branch. The paradigm of the "counselor to the king" (in many cases the king's advisor had complete control of the king simply by manipulating the information input to him) is one which suggests the potential dangers.

with the group of trustees, directors, or whatever might be the designation of the publicly responsible principals of the evaluative branch, then there is basically very little difference between this concept and the Commission's recommendation—the key functions of the new branch and of the Commission's body . . . being the same. If this line of reasoning is sound, then most if not all of Daniel Bell's arguments supporting the use of commissions in government also become reasons in favor of the new branch.

To help insure public understanding of the functions and constraints on the new branch, as well as to assure its stature and independence as a new, fourth branch of government, it would seem best that it be established through the difficult, uncertain, prolonged, but very public process of constitutional amendment.

In view of the apparent and amply publicized defects of existing branches, there is evidently no guarantee, with or without the constitutional amendment route to its establishment, that the evaluative branch would eventually emerge as the very model of a governmental agency. On the other hand, within the inescapable limitations of multi-branch government, there is no reason to believe that the new branch would be unable to achieve in time the prestige, the relative functional independence, and the influence for progress now possessed by other branches—but with the added gain that such achievements would be in increasingly important functions now being largely nonperformed. The point here, simply, is that there is no reason whatever to believe that the new branch, in time, will not do the needed job at least to the level of general performance demonstrated historically by other branches.

The eventual detailed character and usefulness of the new branch would, of course, be determined to a greater extent by the quality of people who would do its work than by any other element. Evidently, one of the main objectives in defining its initial characteristics and prerogatives by law would be to help insure that: (1) its staff members would be chosen from among the leading analytical minds in the professions, business, and the social and physical sciences; (2) they be provided with a politically neutral environment comparable to that of the major universities, and with conditions of employment assuring comparable degrees of continued intellectual vitality and independence; (3) salary levels fully competitive with those paid for equivalent ability and experience elsewhere in our society, and providing within the government as a whole at least the relative economic status and career stability now enjoyed by the judicial branch. Such conditions suggest that the methods of selection, employment, and compensation for the proposed branch would differ markedly from those now being generally used, for example, in the federal establishment. The facts that the branch would be wholly new and substantially independent of the others should enable taking advantage of the positive aspects of experience to date with the civil service system without carry-over of its handicaps.

In the absence of a reasonably specific organizational and functional model for the evaluative branch little can be gained from discussion of the possible details of its charter, numbers of personnel, size of budget, legal assurances, or limitations on continuity of appropriations in case of dissatisfaction with its work by other branches, methods of selecting, appointing or electing (and removing) the trustees and the principal executives, and so on. However, there may be value in clarifying the nature and implications of the concept by suggesting several of its possible characteristics whose persistence could be assured independently of different choices for details of method of establishment, internal organization, general character of interfaces with other branches, and so on. For example:

(a) Since the evaluative branch would have no responsibilities other than collecting pertinent data, analyzing its implications, developing alternatives and their cost/effectiveness characteristics, and evaluating and reporting the performance of other branches of government *vis à vis* goals they had themselves established, or those that had been defined for them, and because it would be as relatively independent, permanent, and as well established by law as the judicial branch, there would exist no rational incentive for its staff to limit or misinterpret the implications of analyses, or to otherwise adopt any point of view other than that of maximizing the accuracy of its estimates, predictions, and evaluations. In these respects, the branch would have the same motivational characteristics that now prevail in the inquiries and researches of the academic community in which the central tests of value are relevance and truth.

(b) As our society has grown more complex, it has become increasingly difficult for the individual citizen to understand the problems and alternatives which he must seek to understand and resolve through the available political machinery. This result is a natural consequence of the increasing number of sources of data and the growing volume, variety and complexity of information to which the citizen is exposed and is required to use. Thus, the task of the citizen in exercising his public functions intelligently has become incomparably more difficult in recent years and shows every prospect of becoming more so. Moreover, the legislative and executive branches at all levels of government have not been able to ease significantly their own or citizens' information problems. Hence one of the most important and useful functions of the evaluative branch would be to develop and coordinate (but not necessarily to operate) national, state, and local information collection and processing systems aimed at providing the main agencies of government at all levels and the public with objective and timely information concerning status, changes, and trends in the key characteristics of our society. The availability of this service to the public, in addition to its direct benefits to government at all levels, would provide a means for curtailing the effects of public misinformation and confusion in many areas. At the least, it would provide a politically neutral source for resolving the numerous controversies on important issues inside and outside the government which now can go in lieu of action because pertinent data is unavailable, or cannot be relied on because it has been selected to reflect a particular bias, or has been manipulated to help substantiate a conclusion which more professional analysis might prove unwarranted.

(c) Since the branch would not be subject to domination by

the executive, the legislature would be likely to employ its resources to secure the information and analyses it needs or desires. Because, aside from public hearings, the legislative branch has no independent source for responding to its more elaborate and searching inquiries, the new branch could fill what now tends to be an information and idea vacuum in many of the complex issues which legislatures at all levels of government are now required to face and overcome.

(d) Aside from providing the data gathering and analytical capability for assisting the executive branch in coping with the major long-range problems concerning our society, the evaluative branch would have the organizational and technical capability, as well as the independence needed, to undertake the objective studies for support of the shorter range objectives of the Planning, Programming, Budgeting System and of its eventual successors. Thus, the evaluative branch could provide the administratively unbiased program analyses and program memoranda needed by the PPBS for evaluating and/or authenticating the near term (i.e., those for the next one to four years) budgetary intentions and plans of agencies in the federal executive branch. Thus the PPBS would be enabled to perform most of the functions at the agency level recommended by Schultze and would complement the work of the evaluative branch in focusing on the interagency, multi-program, multi-year, national level evaluation and development of alternatives. The extension of both the PPBS and evaluative branch activities to the lower levels of government, with the private sector fully informed, would help establish the factual and analytical basis for self-consistent, quasi-consultative planning by both the governmental and private sectors, the importance of which has been so strongly emphasized by Colm and Gulick.

(e) It is clear that the social sciences would be the most heavily represented specialties of personnel of the evaluative branch. Since the branch would seek to develop key data and analyses important for the definition, prediction, and eventual resolution of the central problems of our society, it would provide a basis for fruitful and rapid development of the social sciences.

(f) One of the shortcomings of present federal arrangements for society's management is the absence of a natural location within the executive branch for the central handling of scientific, technological, economic, and other social data with the view of optimizing the value of information outputs to the national system as a whole. An evaluative branch would be the natural locus for a national economic and technical data bank, for data processing, and for information dissemination to the main agencies of government at various levels, as well as to private business and the public. Direct involvement of social scientists with pertinent current data covering essentially the entire socioeconomic system, with the short and longer term forecasting of major courses of change, and with the establishment of data collection and analytical systems designed specifically for the purpose of evaluating prior predictions and improving techniques for these purposes—activities all of which are basic to the long run effectiveness of the evaluative branch—would seem to provide an unequaled opportunity for the social sciences to approach the use of the experimental methods of the physical sciences as closely as it will perhaps ever be possible for them to do.

In essence, then, the evaluative branch would be a national information generator providing guidance for national, regional, and local planning and evaluation, but from a position *outside* the existing power structure, and one *not having* any substantive operational responsibilities in the system. Thus, neither through its establishment nor through its activities after establishment, would the existing shares of powers among the key institutions in the system be significantly disturbed. On the other hand, the information generated by the branch would be directly pertinent to the effectiveness of all institutions and, because of its neutrality with respect to their program postures and bargaining positions, would provide the *enabling* basis for concerted institutional behavior in social and other major problem areas. Such *enabling* would be analogous to the process of catalysis in chemical reactions in which "the reaction between two or more substances is influenced by the presence of a third substance (catalyst) which remains unchanged in the process." [22] In this analogy, the evaluative branch could provide the data, the analyses, the alternatives which would catalyze the reactions between the public (including the Fourth Estate), the business community, and the executive, legislative, and judicial branches. A more pertinent and durable consensus, and one more heavily weighted by considerations of general welfare than of special interests, is likely to result from adversary encounters and bargaining interactions catalyzed in this way. Its status as a constitutionally established, independent branch would help insure that the evaluative branch remained unchanged in the process!

Finally, it is necessary to note that with appropriate but relatively obvious changes the entire argument can be applied to the management of large enterprises of any kind, other forms of government or corporate businesses, for example. The basic idea is that every large enterprise should clearly separate responsibilities and functions concerned with data gathering, analysis of alternative strategies, and the evaluation of organizational performance from responsibilities for short term or tactical planning and the conduct of operations. In the absence of such separation realistic, self-adaptive behavior is likely to be thwarted or strongly inhibited through the inertia of prior commitments, special managerial interests, inability to recognize and/or admit errors, and so on. In the case of a large corporate business, for example, the argument would suggest that a high quality, high status, analytical, and evaluative staff responsible only to the board of directors, and in no way subservient to general management, should in the long run contribute greatly to the success of the enterprise.

I have benefited greatly in developing the above discussion through criticism of earlier drafts by many friends and associates. I want to acknowledge especially the stimulus of suggestive questioning and comment from Harvey Brooks. Others whose comments have been notably helpful are: Joseph Doherty, James Fletcher, John Ford, William Hooper, Jon Miller, Elton Mayo, John Rothberg, Henry Rowen, Russell Smith, and James Webb.

This draft is based on earlier drafts dated August 4, 1967 and April 2, 1968, prepared while the writer was a staff member in the Office of Science and Technology, Executive Office of the President. It is the partial result of work now being supported by the Ford Foundation and including assistance from the Brookings Institution and Harvard University (The John F. Kennedy School of Government and the Division of Engineering and Applied Physics).

REFERENCES

1. H.R. 12291, 90th Congress, 1st Session.
2. *Report of the National Commission on Technology, Automation, and Economic Progress, 1,* February 1966.
3. N. E. Golovin, "Retrospect on AIAA/ORSA Forum." *Astronautics and Aeronautics,* June 1968.
4. Harvard University, *Program on Technology and Society, First Annual Report of the Executive Director,* November 1965.
5. E. G. Mesthene. "How Technology Will Shape the Future." *Science, 161,* July 12, 1968, pp. 135-143.
6. *Webster's New Collegiate Dictionary.*
7. A monograph by Charles Schultze covering the subject matter of the 1968 Gaither Lectures will be published later this year by the Brookings Institution, University of California, Berkeley, April and June 1968.
8. Charles L. Lindblom, "The Science of Muddling Through." *Public Administration Review, 2,* Spring 1959; Decision-Making in Taxation and Expenditures." *Public Finances,* National Bureau of Economic Research: *Needs, Sources, and Utilization,* Princeton, 1961; David Brayboake and Charles Lindblom, *A Strategy for Decision,* New York, 1963.
9. Gerhard Colm and Luther Gulick, "Program Planning and National Goals." To be published by the National Planning Association, Washington, D.C. (Draft received August 1968.)
10. John Hackett and Anne-Marie Hackett, *Economic Planning in France* (Cambridge: Harvard University Press, 1963).
11. John K. Galbraith, *The New Industrial State* (Boston: Houghton Mifflin Company, 1967).
12. *Fortune,* July 1967, pp. 90ff (The reviewer is Irving Kristol).
13. *Business Week,* July 8, 1967, p. 77 (The reviewer is anonymous).
14. Richard E. Neustadt, *Presidential Power* (New York: John Wiley & Sons, 1962), p. 33.
15. Douglass Cater used the phrase, *Fourth Branch of Government,* as the title of a book dealing with the influence of the press on governmental processes. (Boston: Houghton Mifflin Company, 1959). Also, R. G. Tugwell used the term "The Fourth Power" as the title of a

paper in *Planning and Civic Comment*, April-June, 1939. The "fourth power" would be exercised by an agency independent of the rest of the executive branch, and concerned with *direction* of the national economy in the interests of maximizing production and optimizing the use of resources.

16. John William Ward, "The Trouble With Higher Education." *The Public Interest,* Summer 1966, p. 76.
17. Glenn T. Seaborg, "The University in a Changing Society." Address at Howard University, Washington, D.C., March 1, 1967.
18. *Carnegie Corporation 62nd Annual Report, New York Times,* November 20, 1967, copyright 1967 by The New York Times Company. Reprinted by permission.
19. Harvey Brooks, "Applied Science and Technological Progress," *Science 156,* June 30, 1967, p. 1706.
20. Aaron Wildavsky, "The Political Economy of Efficiency." *The Public Interest,* Summer 1967, p. 34.
21. Daniel Bell, "Government by Commission." *The Public Interest,* Spring 1966.
22. *Van Nostrand's Scientific Encyclopedia,* 2nd Ed. (New York: D. Van Nostrand Company, Inc., 1947), p. 260.

Men and Machines in Air Traffic Control

J. W. RABB

Federal Aviation Administration
Washington, D.C.

The Federal Aviation Administration (FAA) of the U.S. Department of Transportation is responsible for the safe and efficient management of all air traffic within the continental United States. The total system of air transportation has many segments such as En Route Air Traffic Control, Terminal Air Traffic Control, Airports, Navigational Aids, etc. This paper discusses only En Route Air Traffic Control and the program to apply automation to this segment of the total system.

An organizational entity (The National Airspace System Program Office) has been established within FAA with the authority and resources necessary to execute: A program to design, specify, procure, install, test, activate, document, and support an equipment and computer program system and to train personnel to provide a first step of automation to En Route Air Traffic Control.

The En Route Air Traffic Control (ATC) system in use today in twenty Air Route Traffic Control Centers (ARTCC's) can justly be classified as a manual system in that the principal elements are human and the tools they use are not automatic. The program defined above is intended to convert this manual system to a collaborative system consisting of human operators using some automatic tools.

The general pattern of air traffic in the continental United States is shown in Fig. 1. As the figure indicates, the heaviest traffic concentrations are in the Northeast (roughly bounded by Washington, Chicago, and Boston), transcontinental between Northeast and Southwest, and up and down the East and West Coasts. This traffic is controlled from twenty ARTCC's with geographic boundaries as shown in Fig. 2. This figure also shows the location of all of the radars presently in operation or planned for the system.

A thorough discussion of the methods and procedures of air traffic control is not germane to the purpose of this paper. Accept, if you will, the following statement as fact: In order to perform his assigned tasks, an air traffic controller needs at least the following information and aids; (a) the position in three dimensions of all aircraft within his sector of jurisdiction; (b) the identity of each of the aircraft within his sector that is under his control (his sector may contain a mixture of controlled and uncontrolled aircraft); (c) the pilot's intended course of flight for the controlled aircraft; (d) facilities to coordinate with other controllers (hand-off of control jurisdiction is one of the important tasks) ; (e) communication with the pilots of controlled flights to give control and hand-off instructions.

A gross pictorial diagram of the manual system used in each of the twenty ARTCCs (now partially replaced in some centers) is shown in Fig. 3. As shown on this diagram, the radar controller, who is responsible for all air traffic in a specific geographical area called a sector, acquires and/or gives information through four separate channels.

First: The positions of aircraft within his sector and a limited amount of identifying information are obtained from a col-located radar/beacon site usually located some distance from the center building. Transmission of this information from the

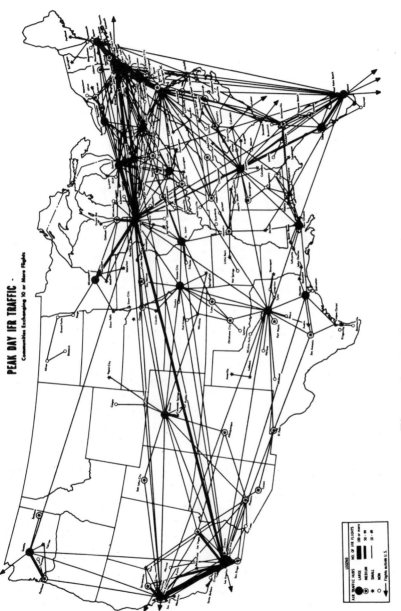

Figure 1.   Peak day IFR traffic.

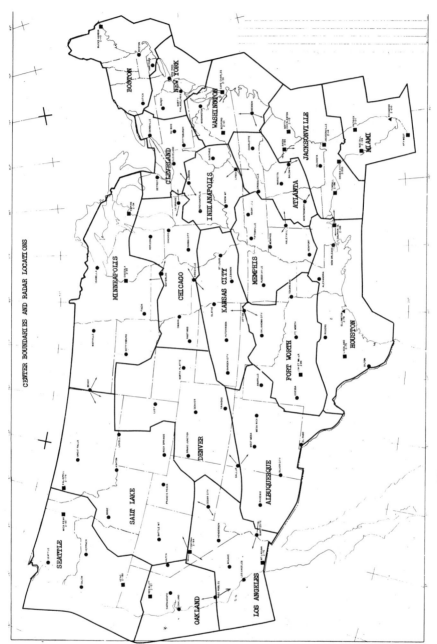

Figure 2. Center boundaries and radar locations.

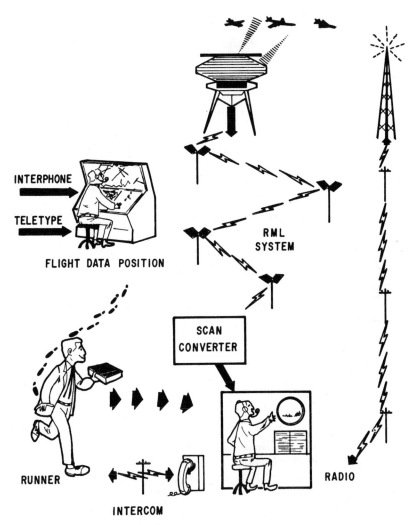

Figure 3.   Manual radar system.

radar site to the center building, and transmission of control signals from the center to the radar site, requires a total of four microwave communication channels of six MHz bandwidths each. By this means, each controller is provided a radar display of aircraft within his control sector.

Second: The pilot's intended route of flight (the flight plan) is received at the ARTCC by voice or teletype communication circuits. This flight plan information is recorded on a strip of paper, mounted in a plastic holder, hand carried by a runner to the control position, and mounted in a tabular bay for display to the controller.

Third: A telephone intercommunications system is provided to connect each control position with every other control position with which that position must communicate for any purpose, such as to transfer control jurisdiction.

Fourth: Two-way radio telephone channels are provided between the controller and all pilots within each sector. By this means, a controller talks with pilots to obtain aircraft altitude and positive identity and to transmit control instructions and other information.

This manual system is being replaced with the partially automated system shown in Fig. 4. This system can justly be classed as collaborative. The same information-communication channels are provided as in the manual system, but a degree of automation has been included in two of them, the radar channel and the flight data channel. Although the intercommunication channels and the radio telephone channels remain, their usage will decrease because some of their functions are transferred to the automated channels.

Radar position information and radar beacon information (on aircraft equipped with an appropriate transponder) are converted to digital form at the radar sites by digitizers developed jointly by the FAA and the Air Force. One digital message is composed for each aircraft within the radar field of view on each antenna rotation. These messages contain range and bearing for all aircraft and, in addition, altitude and identity for those aircraft that carry appropriate beacon transponders. The messages are routed to the FAA, ARTCC, and, where required, to Air Defense Command facilities. The transmission medium from each radar site to the ARTCC building is a three channel system of 2400 bits per second per channel.

In the center building, the radar information in digital form enters a large digital Central Computer Complex (CCC). Here it is processed and information is prepared for display to the controllers via a Computer Display Channel (CDC) to provide

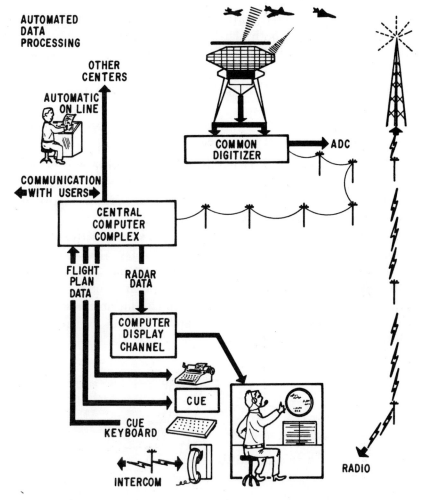

Figure 4. NAS en route stage A system.

the following broad categories of air traffic control functions:

1. Automatically and manually initiated computer program tracking of aircraft position.
2. Bright display of alphanumeric and radar data.
3. Entry and processing of flight plan information.
4. Flight progress strip printing at appropriate sector positions.
5. Provision for entering and receiving new and revised flight data (updates) at all operating positions.

6. Intersector coordination through computer generated alphanumeric displays, both plan view and tabular.
7. Interfacility coordination through the use of computer transmitted data (computer to computer).
8. Computer generated displays of geographic (map) and weather data.
9. Automated computer initiated hand-off of control jurisdiction with manual override.

There are human interfaces with the system in several places but the two principal areas of real-time on-line are the radar air traffic control positions and the aircraft pilots. Figure 5 shows the principal features of a radar controller console, a part of the CDC. It contains a twenty-three inch Cathode Ray Tube (CRT) on which a Plan View Display (PVD) of the controller's area of interest is generated by the CCC. It also contains a small CRT as a part of a Computer Readout Device on which the computer can write alphanumeric messages to the controller. In addition, there are numerous controls, indicators, and data entry devices.

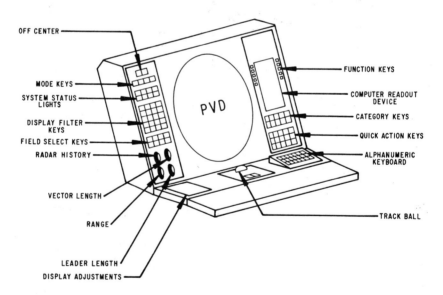

Figure 5.   Radar controller console.

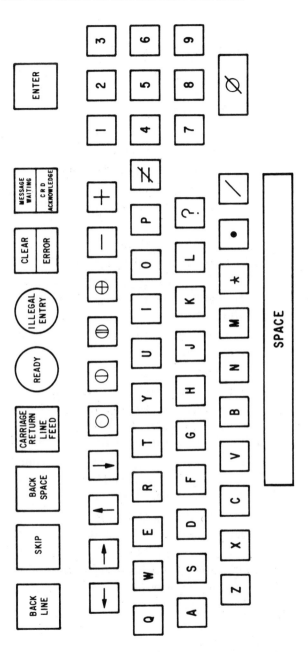

Figure 6.   Alphanumeric keyboard.

Special Symbol	*Meaning*
1. ◇	Flight-plan-aided track position symbol
2. ◁	Free track position symbol
3. #	Coast track position symbol, or an entry in a hold list is not at a posted fix
4. □	Flight plan position symbol
5. •	Uncorrelated primary radar data
6. x	Correlated primary radar data
7. /	Uncorrelated beacon radar data; B_4 character in full data block—aircraft's assigned altitude is "OTP"
8. \	Correlated beacon radar data
9. ≡	Beacon radar data containing SPI (Ident)
10. ⌐	Common Digitizer map symbol
11. ↑	B_4 character in full data block—aircraft climbing
12. ↓	B_4 character in full data block—aircraft descending
13. +	B_4 character in full data block—aircraft out of conformance limit (high)
14. —	B_4 character in full data block—aircraft out of conformance limit (low)
15. □	Trackball position marker
16. ?	The expect further clearance time for an aircraft in a hold list is within 5 minutes of the present clock time.
17. ⌃ (accent)	The flight plan position of a tracked aircraft has passed the first posted fix in a sector without an Accept Handoff action. The accent character will appear over the first character (A_1) of the aircraft identification.
18. →	Not assigned
19. ←	Not assigned
20. *	Not assigned

Figure 7.   Display symbols.

The controller can transmit a complete set of alphanumeric characters and some symbols (see Fig. 6) to the computer and the computer can display the same set of characters and symbols on the computer readout device. The computer makes up the

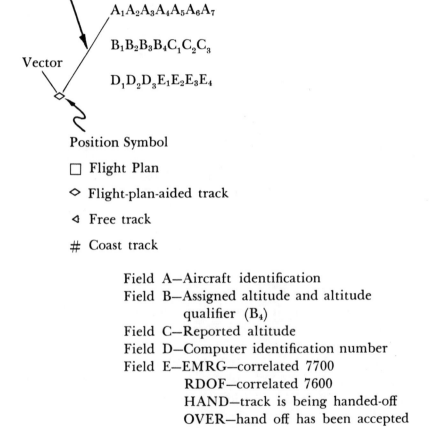

Leader

$A_1A_2A_3A_4A_5A_6A_7$

$B_1B_2B_3B_4C_1C_2C_3$

Vector

$D_1D_2D_3E_1E_2E_3E_4$

Position Symbol

□ Flight Plan

◇ Flight-plan-aided track

◁ Free track

\# Coast track

> Field A—Aircraft identification
> Field B—Assigned altitude and altitude
>   qualifier (B_4)
> Field C—Reported altitude
> Field D—Computer identification number
> Field E—EMRG—correlated 7700
>   RDOF—correlated 7600
>   HAND—track is being handed-off
>   OVER—hand off has been accepted
>   Established beacon code

Figure 8.   Full data block format.

PVD using these same characters and symbols, the special symbols listed in Fig. 7, and straight line segments.

Each aircraft of interest to a controller can be tracked by the computer. A data block (see Fig. 8) is associated with the air-craft position on the PVD. Any, all, or none of the fields of

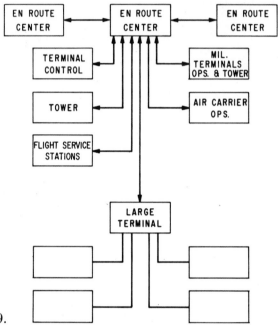

Figure 9.

information shown in the figure can be selected for view by the controller. Certain information, such as EMRG, RDOF, HAND, and OVER in Field E, is forced on the display and cannot be de-selected.

The nationwide system will be interconnected by digital data links as shown in Fig. 9. Each ARTCC will connect with every other ARTCC with which it shares a common boundary. It will also connect with other facilities within its boundary as shown.

In total, it is a very large and complex system of machines, people, computer programs, and procedures. Its operation is highly dependent upon people such as controllers, maintenance personnel, management personnel, pilots, aircraft owners, pas-sengers, members of Congress, and many others with direct or casual interest. These act as individuals and/or members of social organizations. In general, the only group of individuals involved that is not represented by an articulate organization is the passenger group.

If one considers the physical size, the technical complexity, and the sociological intricacies of this national system, it is easy to see that the fulfillment of the FAA mission of air traffic control is difficult and expensive. Multi-disciplinary problems of significant magnitude exist. Every known method for problem solving is being applied or being considered for application to these problems.

One very useful tool that has been used extensively is simulation or modeling. The problems and pitfalls of this technique are well known. The following are perhaps the two most troublesome in this case. 1. The elimination of prejudgment or bias on the part of the model designers is difficult because of their knowledge and background in the present manual system. 2. The human components, especially the controllers and the pilots, are critical elements and almost impossible to adequately simulate.

To minimize the first problem, a great deal of industrial and institutional talent has been obtained under contract. The second problem, obviously, is much more difficult, as might be expected; the most successful simulations have used human operators to simulate human operators. Even so, the actions and reactions of the operators in a simulation environment are quite different from those in a real life environment.

Realism in simulation is extremely important, especially in the most critical elements of the system being simulated. Therefore, in the simulation of systems such as the one briefly described herein, there exists a certain dichotomy. Since the systems are large and complex, and since they involve many people and perhaps life-or-death situations, it is highly desirable to employ simulation as one of the problem solving techniques. On the other hand since the human is a critical element and impossible to faithfully simulate, the results of such an effort can be misleading or completely invalid and certainly should be thoroughly analyzed before final decisions are made based upon these results.

In summary, solving the air traffic control problem is difficult and expensive. Every known technique for problem solving on a multidiscipline basis is being employed or considered. One of the most used to date is simulation or modeling. The results of these simulation efforts are of limited value since large numbers of people are a part of the system being simulated and people cannot be adequately simulated. To those of you who would simulate large social systems involving hundreds, thousands, or possibly millions of people, I can say that you have a very difficult task and I can really offer little except sympathy.

REFERENCES

1. T. V. Garceau, "Definition of the Computer Display Channel (CDC) Console Controls." MITRE Technical Report—4027, NTR #84 DE, Revision 1, June 21, 1967.
2. "National Airspace System (NAS) Fact Book." Air Traffic Training NAS Series, FAA Aeronautical Center, February 1, 1968.
3. "System Description National Airspace System Enroute Stage A." SPO-MD-109, System Program Office Configuration Management Directive, FAA, May 30, 1968.

Complex Interrelations Affecting the Reliability of Bulk Power Supply

F. STEWART BROWN

Federal Power Commission
Washington, D.C.

In the Federal Power Commission's report on the *Prevention of Power Failures*,[1] the term reliability was defined as meaning, "the ability of a utility system or group of systems to maintain the supply of power. Reliability is gauged by the infrequency of interruptions, the size of the area affected, and the quickness with which the bulk power supply is restored if interrupted."

The Commission's report listed some thirty-four recommendations and conclusions relating to the reliability of bulk power supply. These include recommendations for strengthening coordinating organizations on both a regional and national basis, the building of stronger transmission interconnections in many areas, extending the lead time in the planning of new facilities to avoid defaults in initial operating times, the wider use of automatic load shedding as an insurance factor against total failure, and numerous provisions relating to controls and emergency facilities. Among the latter, the Commission recommended that "utilities should intensify the pursuit of all opportunities to expand the effective use of computers in power system planning and operation." The report also observed that many new applications are being found for high-speed digital computers in the planning and operation of utility systems.

Electric power in the United States is supplied by some 3500 systems of diverse size and ownership. About ninety-five per cent of the supply comes from 200 of the larger utilities. The great majority of all systems are interconnected in a transmission network which now spans the forty-eight contiguous states, except that most of the utilities in Texas, while interconnected among themselves, do not have synchronous ties with surrounding utilities. Figure 1 illustrates the progression since 1939 of interconnection and synchronous operation of utility systems.

Power demands in the United States have been growing for several decades at an average rate of about seven per cent a year. This means a doubling in the generating capacity every ten years to meet these demands. The generating capacity of all electric utilities in the United States now stands at about 280 million kilowatts; including industrial generating plants, the total is nearly 300 million kilowatts.

Failures in bulk power supply stem from a wide variety of causes. The Commission collects and publishes data on all interruptions which result in a loss of 25,000 kilowatts, or more than one-half of the system load, for fifteen minutes or longer, and which involve facilities operating at 69 kv or higher voltages. Of the forty-seven power interruptions reported to the Commission during the first six months of 1968, twelve were caused by natural phenomena, mostly wind combined with snow and icing conditions during the early part of the period, and lightning during the latter part of the period; twelve originated from equipment failures, primarily conductor and insulator failures; and eight were caused by improper operations, mainly errors in relay wiring and a few operator misjudgments. Of the remaining fifteen outages the initiating cause was unknown in six instances, four were due to gunshot, three to damage by aircraft and two to other causes.

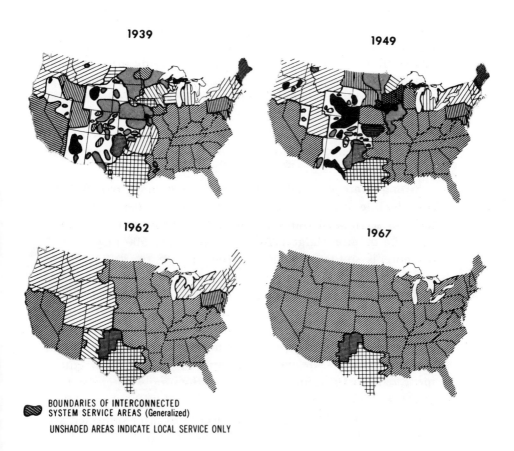

1939

1949

1962

1967

BOUNDARIES OF INTERCONNECTED
SYSTEM SERVICE AREAS (Generalized)

UNSHADED AREAS INDICATE LOCAL SERVICE ONLY

Figure 1. Growth of interconnected systems operating in parallel.

Regardless of the origin, any precipitating cause can lead to a major failure unless the interconnected network is of adequate strength and is equipped with properly functioning protective and control devices. The widespread power failure in the Northeast of November 1965 started from the operation of a transmission line relay that had been set too low. Widespread failures usually involve system or area separation from the network. Most times the isolated area is left with a deficiency in the supply of generation in relation to load demands. If quick action is not taken to relieve part of the load demand, the alternating current frequency will decline rapidly, and early total collapse of power supply is likely in the isolated area.

Strong interconnected alternating current power systems have considerable capability of self-preservation when generating units or sections of transmission lines suddenly fail. It is likely that one or more large generators producing 500 megawatts * or more is suddenly removed from service by its protective equipment several times a day somewhere in the United States without affecting power service. This is one of the great virtues of strongly interconnected operation. Another major benefit comes from the ability to move large blocks of power from one area to another to supply a deficiency caused by unusual weather extremes, delays in the initial start-up of new equipment, damage to an important part of the supply system, strikes, or other contingencies.

Major economic gains also can accrue to interconnected systems through coordinating the planning and operation of large facilities, taking advantage of diversity in loads and employing the most economic units available in an area for energy supply.

Interconnected alternating current networks possess inherent characteristics, derived from the massive rotating momentum of the many electrically interlocked generating machines, which aid in supporting the synchronous integrity of the network. If a generating unit suddenly fails, power immediately flows into the area from all generators on the interconnected network. The flow of a.c. power is controlled by the relative angular posi-

---

* A megawatt is 1000 kilowatts.

tions of the machine shafts at different locations in the network. If the sudden demand for power is large, the in-rush may exceed the present limits of protective transmission line relays; or the shift in machine angles may exceed limits within which the flow would continue to increase, resulting in an out-of-phase relation which requires opening of the circuit to avoid possible generator damage. Such line openings by overload or instability are a common occurrence in widespread failures.

In all of this great complexity of system planning and operation analogue and digital controls are extensively applied. Many of these devices perform a protective function to prevent damage to a piece of equipment. Their action usually results in the removal of an item of equipment from the operating position. If, for example, lightning strikes a transmission line and creates an arc path from phase to ground or between phases, a protective relay will quickly activate the opening of a circuit breaker to de-energize the line and extinguish the arc. Because such strikes usually cause little damage, automatic line reclosing is a part of the control program of most utilities. Opening customarily takes place within three or four cycles following the lightning stroke and reclosing about twenty cycles later, the time usually required for decay of the ionized arc path.

In addition to relays which detect abnormal power characteristics, others perform such functions as the detection of gas in high voltage transformers or excessive temperatures in generating machines. These relays remove the equipment from service or transmit a warning signal to operators. A utility which serves a peak load of three million kilowatts, approximately the power requirement for Washington, D.C. and its surrounding suburbs, possibly would utilize several thousand relays of various kinds on its bulk power supply system. The proper servicing of these relays has an important bearing on the reliability of the utility's operation.

Current and voltage and their associated phase relationship are the basic electrical characteristics which are sensed and evaluated to control and protect network operations. As one can imagine, the variety of relays that can be devised utilizing these basic determinants in combination with time variants and integrations is practically unlimited.

It may be useful to note several examples of relays which have uniquely discerning characteristics. One relay assembly utilizes voltage, current, and time to make rate of change measurements and thereby discriminates between line faults and out-of-step conditions which may have similar characteristics but may warrant different control actions. For example, when a true fault occurs, the relay triggers the immediate opening of a circuit breaker on the line. If a power swing occurs the relay may prevent opening or delay the reclosing of the circuit, depending upon the desired control actions for particular operating conditions. Another example of a relay possessing a highly selective sense of discrimination based on voltage, current, and phase relationships is the so-called mho-type relay. It basically measures circuit admittance and is sensitive to different angular relationships between the operating voltage and currents so that it will operate to trip its terminal circuit breaker for an abnormal condition that may involve voltage and current magnitudes no different from those present under normal operation, but which possess a different phase relationship because of a short circuit or other abnormal operation condition. Although the art and science of instrumentation and control in utility practices is highly developed, it is still undergoing continual improvement.

Most utilities employ analogue or digital computers to aid in the dispatching and control of power flows. It has become relatively common for a utility to control its power dispatch by a computer which continually analyzes the load and supply situation and, for the conditions at any time, selects the increment or decrement of generation which is economically most advantageous. Computer programs for this purpose can become rather complex, particularly if a system has a combination of hydroelectric and thermal energy supply. The more elaborate programs include the evaluation by computer of transmission line losses in addition to the economy of plant generation.

While many economic dispatch computers operate on-line,

they do not respond instantaneously to major system disturbances. Flows between systems are regulated to agreed upon rates of exchange, which may include many variations during the day. Interconnected utilities set the rate of response of their generating machines in terms of an agreed upon tieline bias. Frequency bias settings are in terms of megawatts of increase or decrease in system generation per unit change (usually 0.1 Hertz) in frequency of the interconnected network. Controls to accomplish this are relatively simple and tend to avoid the hunting in tieline flows that were troublesome in the early days of interconnected system operation. However, tieline bias is not intended to motivate, and is not capable of motivating, a quick response to sudden emergency situations.

Utilities generally agree that the most practical and effective employment of computers for the immediate and early future is in the field of more complete and more timely reporting and analysis of system operating conditions, including warning of situations which are developing or have exceeded preselected operating limitations. Also, it is of substantial assistance to system operators to run quick system security checks to determine what actions may be advisable in shifting the generation or making other system adjustments if it appears that elements of the network may be approaching capacity limitations.

System security analyses often employ distribution factors of varying complexity and refinement which have been programmed in the computer. Such analyses do not place any undue demands upon computer capacity or performance, but they do involve a very substantial amount of off-line network analysis and programming. Because of the continual growth of power demands and required generation and transmission additions, updating of the program must be a relatively continuous process.

A group of utilities in the eastern part of the country, which probably has advanced as far as any in the application of digital computers to network control, employs an IBM 360/50 computer which is programmed to provide for monitoring system security, the economic dispatch of all units contributing to the pool operation, continual evaluation of generation reserves, the monitoring of megawatt flows on principal transmission lines, the determination of pool-to-pool schedules at hourly intervals, and the establishment of master load and generation schedules four times daily. The computer receives information from 255 points on the transmission network, representing about one-third of all lines of the interconnected system. Analogue information is collected at a selected number of remote terminal stations where it is converted to digital signals and transmitted to the computer at the master terminal. The master terminal scans and receives signals at the rate of 600 bits per second.

Bonneville Power Administration has been conducting a large-scale computer application study. While the functions they have in mind, I believe, are not greatly different from those I have just described, the proposed application would include compiling and processing data for the entire West Coast. Thus, the magnitude of the job is tremendous. It would also be necessary for the installation to be able to communicate via computer with the several varied installations now in different degrees of development by the individual utility systems on the West Coast.

Analyzing the transient stability of a network subjected to a severe disturbance requires much more computation than steady state flow determinations. A satisfactory analysis requires representation of networks on a regional pattern plus some representation of the network beyond regional boundaries. In this area I think we are beginning to approach the limits of capability of the third generation computers. Utilities in the west, for example, have recently completed programming for stability studies of a 1400 bus system with 400 generators. The program includes provision for governor characteristics and variation in load representation with varying system conditions. The program is written for an IBM 360/65 with 512,000 bytes of core storage. This, as you may know, while not the largest, is a fairly respectable machine. In a stability analysis this machine requires about one hour to duplicate the real-time performance of the power system over about three seconds. This calculation takes account of the energy transfer between machines and the varying power flows over the lines as the machines swing back and forth with respect to each other until an energy balance is restored, if the system is stable, or until the power flow exceeds the line transfer capabilities or the relay settings, causing the system to break up because it is unstable under the postulated disturbance.

Proper load representation in transient stability studies is a difficult problem. It is not that engineers do not understand how a particular load changes with changes in system frequency and voltage, but they don't know exactly what types of loads the customers have connected to the systems at any particular location and any instant. For example, if the voltage is reduced on an induction motor such as may be used in a refrigerator or air conditioner, the motor draws a correspondingly greater current, so that the electric power taken by the motor changes very little. In other words, the motor load is a relatively constant power load over the voltage range at which it is capable of continuing operation. An electric water heater, on the other hand, will take less current and less power as the voltage is reduced; it is essentially a constant impedance load.

As a means of maintaining voltage, utilities often install capacitors which, on the system, look like capacitive reactance loads. These capacitors help hold up the voltage on long feeders. As system voltage falls during a disturbance, however, these devices take less leading current and allow the voltage on the feeder lines to fall even faster.

Thus, the correct representation of load on a substation requires a mixture of elements which have constant power characteristics, others that are constant impedance, or still others that may be best represented as constant current. The trouble is that we cannot be sure just what the load mix is on a general service substation, so that we cannot be sure we have the appropriate representation.

It is difficult to foresee at this time when or in what manner digital systems may be developed which could substantially take over the analysis and control of system operation in real-time, including control during severely disturbing situations on a.c. systems. Such total programs of control are challenged by the sheer magnitude and extreme speed of the sensing and communication requirements and the programming for a basically complex network under an almost unlimited variety of eventualities. The redundancy of elements of the total control system to avoid misoperation or non-operation could be most demanding. Such an undertaking would far surpass, to my knowledge, any system of digital analysis and control now in operation or planned.

For about the past year and a half, IBM has been making a study for the Public Service Commission of the State of New York of what might be accomplished in the way of advanced computer control of power systems. The company's initial efforts have been to develop mathematical models which would simulate the operation of various components, or modules, of a power system.

Models formulated include an electric generator, exciter, and voltage regulator; hydro plant; fossil-fueled boiler and steam turbine; nuclear-fueled steam generator and turbine; protective relaying; load-frequency control; appropriate load representations; and phase shifting transformers. These models can be combined to represent a specific system. Communication and terminal requirements for substations and generating plants have been examined, including the possible application of small computers at substations. Dispatch procedures and facilities have been analyzed, including the feasibility of on-line load flow analysis and information display systems including CRT presentations for operators.

The feasibility of on-line transient stability analyses is also being examined. It is likely that any significant progress here in the near future will require a very simplified representation by means of equivalents of large portions of the power system.

Work is also underway, largely dependent on progress on items already described, on the possibilities of using the computer to direct corrective action. This will involve consideration of possible ways of restoring the system to equilibrium after an upset, and means of implementing any such ways via the computer. This would seem to require very advanced sensing, communication, and computing facilities. A report is expected.

It is interesting to consider what other steps may be developed and applied to improve the reliability of electric bulk power supply. Equipment manufacturers are contemplating improvements in the performance of control equipment for large generators and turbines. There is room for improvement in the control of generator exciters to prevent unwanted and unwarranted removal of units from service during system disturbances when voltage may fluctuate rapidly for an interim period. Improvements in governor controls and steam turbine valves should enable faster generator output when major loads are suddenly interrupted. Present relays appear highly sophisticated and yet much opportunity remains in the field of anticipatory controls actuated by variable rates of change in power system characteristics.

A very effective way to safeguard against total collapse of power in any area is through the provision of automatic load shedding relays set to disconnect preselected segments of the load at appropriate declinations in operating frequency. The massive Northeast power failure in November 1965, and others which have occurred subsequently, have convinced most utilities of the wisdom of adding this insurance protection. It is not a substitute, however, for planning and installing adequate generation and transmission facilities.

A research program sponsored by the Edison Electric Institute on the interfacing between a.c. and d.c. transmission reveals the likelihood that much beneficial control for some abnormal conditions can be secured in this manner. The flow of power on a d.c. line connecting two points on an a.c. network is not affected by power swings on the a.c. system. The flow of power, however, on the d.c. line can be readily and instantly adjusted by varying the timing of the firing cycle of the a.c.-d.c. converters. There appear to be no technical obstacles to the development of reliable controls which would sense the need for stabilizing assistance in the a.c. system and activate the desired response on the d.c. line. At present the expense of constructing two sets of d.c. converters, one for each end of the d.c. interlink, restricts consideration of d.c. to particularly difficult and important problem areas.

Apart from the technical intricacies of power system planning and operation, there are many other involvements and problems, discussed in the following:

Interconnected utilities are finding it necessary to build stronger coordinating mechanisms. No utility of an interconnected network can plan and operate its systems without affecting its neighbors in some way. Utilities generally agree that stronger coordination should be introduced at the regional level with overview of coordination between regions on a nationwide basis.

The projection of loads up to seven years in advance is an essential factor in the planning for a reliable power supply. This is a science in itself. The rapidly expanding uses and markets for power, the influence of weather on loads, and the competition of other forms of energy are some of the factors which can make accurate load forecasting an elusive experience.

The impact of utility system construction on our natural resources and environments encompasses the considerations of aesthetics, water quality control, and air quality control. These factors have mushroomed in importance within the last five years.

Delays in manufacturing, in obtaining rights-of-way, in securing adequate labor supplies, and other construction factors must be given due allowance in the programming of new facilities.

Utility planning and operation must track a rapidly moving target, one which is changing not only in quantity but also in quality. A very high order of voltage regulation and the elimination of even momentary interruptions is necessary to avoid disturbing and costly effects for some new applications of electricity. I have already mentioned that new equipment must be installed at a rate which will double every ten years. In some parts of the United States, doubling of power demands is occurring in seven or eight years.

It is difficult to project how soon digital computers will become a principal element in the operating control of inter-connected electric systems. Some of the events and trends in recent years, including a number of major power failures, the growing interdependence of one utility upon another and the demands of power consumers for a higher degree of reliability, have stimulated an interest in the wider application of digital systems for planning and for operating control. It probably will be quite a few years before power systems are largely con-trolled by computers. However, there is good reason to expect that digital systems soon will be widely employed to analyze, display, and warn operators regarding system and intersystem conditions on a highly current basis. Computers should be able to advise operators of appropriate actions to be taken when limiting conditions are approached. If an unusually severe dis-turbance should occur, causing system separation and degrada-tion of supply, computers should be of assistance in portraying the overall situation quickly, in defining the elements that failed or have been interrupted and indicating remedial actions to be taken. Such programs will be helpful in resolving some of the complex interrelations affecting the reliability of the nation's electric power supply.

REFERENCE

1. *Prevention of Power Failures, 1:* A Report to the President by the Federal Power Commission, July 1967. Superintendent of Documents, U.S. Government Printing Office, Washington, D.C.

Modeling, A Fundamental Problem in Air Pollution Control*

JAMES N. PITTS, JR.

University of California

Riverside, California

Students today want and, in fact, demand to be involved in solving relevant problems of our society. "Where's the action?" they ask. Well, there's plenty of action in the highly relevant problem of air pollution where research, development, and control activities have accelerated at a remarkable pace. As Jack McKee, Professor of Environmental Health Engineering at Cal Tech, puts it: "In the sixties, environmental pollution was 'discovered' by three important segments of society, *viz.* journalists, politicians, and women. We [environmental scientists and engineers] need them, and we welcome them to our cause." [1]

Action in air pollution is expressed in many ways. The public is concerned, Congress is concerned, and a large number of local, state, and federal agencies are concerned. They are concerned to the point where, for example, today substantial

---

* This is an abbreviated version and will appear in its entirety elsewhere.

amounts of money are being spent, or contracted to be spent, on the subject of my talk—the development of mathematical models for air pollution control.

Why are such mathematical models important? There are at least two good reasons. First and foremost is that proper models could be invaluable in predicting and preventing air pollution disasters on the local or regional level. Second, with valid models the enormous medico-legal and socioeconomic effects of air pollution can be considered in a rational predictive manner.

I became involved in this modeling problem, and in fact am here today, *not* because I am adept at mathematics or computer technology (my sixth grade daughter, deeply engaged in modern math, feels I'm definitely not with it in these subjects), but because I do know something about air pollution and, specifically, the chemistry of urban atmospheres. Furthermore I recognize that, if reliable and useful models of atmospheric pollution are to be developed, we environmental scientists *must* develop new and strong lines of communication with you mathematicians, scientists, engineers, medical researchers, social scientists, and administrators who constitute the American Society of Cybernetics. The present lack of such communications between members of your society and my professional security blanket, the American Chemical Society, concerns me deeply. We have much in common, including the same initials for the societies (a point that confused our business office no end when I filed a request for a travel advance to go to an *ASC* meeting), but in the past all too often we simply have not communicated.

Let me illustrate with one example the need for such collaboration. Since the early 1950's the Los Angeles Air Pollution Control District has accumulated millions of raw data on concentrations of atmospheric pollutants at various times of the day and in various locations in the Los Angeles Basin. These, together with weather data, would seem ideal for developing diffusion models of atmospheric pollution. Well, the data are there, but on closer examination a chemist could tell you that some of the earlier results are worthless. This was through no fault of the operators but simply because certain of the analytical methods for determining small amounts of pollutants in complex atmospheric mixtures gave, at best, crude and, at worst, wrong results. This is the case for the analyses for nitric oxide (NO) and nitrogen dioxide (NO_2) in the parts per million (ppm) or parts per hundred million (pphm) range. Most of the total oxides of nitrogen (symbolized as NO_x) emitted in auto exhaust is in the form of the colorless, relatively non-toxic NO. But in the presence of hydrocarbons and sunlight this NO is converted rapidly to the highly colored (red-brown), toxic form, NO_2 *(vide infra)*. Obviously the absolute values of the concentrations of NO and NO_2 in the atmosphere must be available if proper diffusion models of reactive smog are to be developed. Well, just keep in mind when writing computer programs for such atmospheric models that the vast amounts of data on NO_2 taken in the 1950's are not valid because the early analytical methods were poor. Recall the old cliche "garbage in, garbage out."

The area of air pollution research and development is particularly appealing to me for several reasons. First, I have been conducting basic research in photochemistry for twenty years and a major class of smog, photochemical air pollution, is initiated and sustained by solar radiation *(vide infra)*. Working in the general area of the chemistry of urban atmospheres permits our group to devote a significant fraction of our research effort to the study of fundamental primary photophysical, photochemical and photooxidation processes of relatively simple molecules under carefully controlled laboratory conditions.

Second, the applied aspects of air pollution research encourage us to become deeply involved in studying key aspects of the highly complex physical, chemical, and biological reaction systems involved in atmospheric pollution. Here, one is concerned with the real world—*what will our results contribute to the overall goal of control of air pollution?* Total involvement here becomes, of necessity, an approach to an understanding of the interactions between large physical, biological, and sociological systems, precisely the topic of this symposium.

Finally, I have a deeply personal reason: my daughter, who is twelve, has intermittent attacks of asthma. Last night at

6:00 P.M., here in Washington, I received a long distance telephone call from our physician in Riverside. Becky had experienced an attack of asthma. He was placing her in an oxygen tent in the *charcoal-filtered*, air-conditioned pediatric ward at the Community Hospital. I asked the doctor what he thought had set it off. He said, "Well, she was fine yesterday, Jim. I see no evidence of inflammation; it's just that the damn smog came in around 3:00 P.M. and that seemed to trigger the attack."

Actually, to the best of my knowledge, there is little, if any, *absolute proof* that photochemical smog initiates asthma attacks. The situation is something like cigarette smoking and heart attacks or lung cancer. It's tough to establish that a definite relationship exists. However, as another of my physician friends put it, "If I had my office at the bottom of Fort Knox and never saw daylight, I could guess the air pollution condition outside by examining my daily attendance list of patients with pulmonary problems."

This isn't just a Los Angeles type of problem. You, here on the Eastern Seaboard, have the classical London type of smog (plus our photochemical air pollution during certain periods in the summer and fall months—we shall discuss *both* types later in this paper). My brother-in-law lives in Baltimore and on cold damp days when the smog level is high (in this case chiefly sulfur dioxide and particulates, *vide infra*) and he must be outside, he takes a tedral pill to ease his throat constriction and assist breathing.

Well, you can see why I have something of a missionary approach to the effective control of air pollution. The problem affects all of us personally, some more acutely than others. It will only be solved by a well-thought-out, multidisciplinary approach, in which an important effort is the development of *effective* mathematical models for predicting local and regional atmospheric pollution levels.

This is well recognized by a variety of organizations, public and private. For example, I have here a "Request for Proposal for Developing and Validating Mathematical Models and Methodologies for Elementary Air Pollutant Concentrations of Various Locations in Any Urban Area." This recent Request for Proposal came from the Coordinating Research Council (CRC) through their Air Pollution Research Advisory Committee (APRAC). CRC funds come from three main sources: the Automobile Manufacturers Association, the American Petroleum Institute, and the Public Health Service through the National Air Pollution Control Administration. This is a rather unique and commendable joint effort.

Today some of the modeling aspects of my talk will be derived from a technical proposal submitted in response to this CRC-APRAC "Request for a Proposal." It represents a joint effort on the part of Meteorology Research, Incorporated of Altadena, California; Systems Development Corporation at Santa Monica, California; and Environmental Resources, Incorporated, Riverside, California. I participated as a member of the ERI team. However, today I am not going to stress the development of a specific mathematical model; instead, in view of the time limitations of this talk, my chief aim is to provide you with certain key facts about air pollution so that you can better understand the complex physical, chemical, biological, and sociological implications of the problem we all face. At the end I shall briefly outline our (MRI, SDC, and ERI) approach to modeling for air pollution control.

Historically, air pollution has been recognized as a major problem of urban man for more than three centuries! Thus in 1661 John Evelyn published *Fumifugium: or The Inconvenience of the Aer and Smoake of London Dissipated. Together with some Remedies humbly proposed.*

Today, as noted earlier, we suffer from two general types of smog: (1) the sooty, sulfurous London smog of John Evelyn and his ancestors, and (2) a comparative newcomer, *photochemical air pollution,* first identified in Los Angeles (circa 1945). It is a product of the age of transportation and special meteorological and geographical conditions. Some areas of the United States suffer from both types, plus additional pollutants.

Table 1 compiled from remarks in Professor Leighton's classic book, *The Photochemistry of Air Pollution,* compares the two types of smog.[2] (Note: For excellent general reviews,

Table 1

Comparison of London Type and Los Angeles Type Smog

London	Los Angeles
Peaks early in a.m.	(Photochemical Air Pollution)
Temperature 30-40°F	Peaks midday
Relative high humidity and fog	Temperature 70-90°F
Radiation or surface inversions	Low relative humidities and clear sky
Chemically reducing atmosphere	Subsidence or overhead inversion
Bronchial irritation	Chemically oxidizing atmosphere
	Eye irritation

see also Refs. 3 and 4.) The typical London smog peaks early in the morning. It occurs at relatively low temperatures and at relatively high humidity. There is a radiation inversion and the atmosphere, which is polluted with particulates and oxides of sulfur, is chemically *reducing*. The most pronounced physiological manifestation of London smog is bronchial irritation.

Conversely, photochemical smog peaks in Los Angeles around noon to one o'clock in the afternoon and characteristically occurs on clear days with high temperatures, low relative humidities, and an overhead inversion or a subsidence. Another fundamental difference is that the atmosphere is chemically oxidizing due to the presence of such compounds as ozone and peroxyacetyl nitrate (PAN) which are formed from the action of sunlight on the complex mixture of hydrocarbons and oxides of nitrogen which, in the Los Angeles Basin, are emitted primarily from auto exhausts.

Photochemical smog is additionally characterized by a variety of other effects. These include severe damage to many types of field crops, ornamental plants and trees, and pine forests; pronounced eye irritation; reduced visibility; cracking of rubber exposed to the atmosphere; and finally, definite loss in vitality of an individual exercising on a smoggy day.

All of these manifestations, plus other more subtle effects, have been observed not only in Los Angeles and Pasadena, but throughout the Los Angeles Basin, including Riverside, sixty miles to the east of Los Angeles, for well over a decade (two decades in Pasadena).

Figure 1 is a relief map of that area with freeways super-

Figure   1.   Relief map of Los Angeles Basin with the freeway system imposed on it.

imposed. This figure shows not only the line sources of our automotive pollution but also the mountain ranges which provide physical barriers to the rapid dissipation of smog. Also shown is the Pacific Ocean which moderates our climate and is a key factor in producing strong atmospheric inversions which place a meteorological lid over the Los Angeles Basin at an altitude of several thousand feet. In part, as a result of this inversion on the all-too-frequent smoggy days, residents of the Los Angeles Basin live in a hazy, eye-smarting, lung-choking atmosphere of aerosols, particulates, and noxious gases.

Plants as well as people are affected by air pollution. The first definite plant damage due to photochemical air pollution was noted by John Middleton and co-workers back as far as 1944. Then in 1951 they showed that the loss of spinach crops in Los Angeles County was due to photochemical smog. It was estimated that the total crop damage was as high as $500,000 in Los Angeles County alone. In the early 1950's, plant damage had spread to the surrounding counties of Riverside, San Bernardino, and Orange, and Haagen-Smit and co-workers had shown the key roles of sunlight and auto exhaust in producing this new kind of smog.[5]

Plant damage from air pollutants and, in particular, photochemical smog is not confined to California. The spread of photochemical oxidants and crop losses of millions of dollars on the Atlantic Seaboard paralleled, on a year-to-year basis, the spread in Southern California, although occurring somewhat later. For example, in 1953 oxidant from photochemical smog was discovered in New York and in 1957 it was identified in Pennsylvania.

Today in New Jersey, the Garden State, pollution damage to at least thirty-six commercial crops has been reported and has been observed in every county.[6] By now, damage by photochemical smog has been identified in most states of the United States and many foreign countries. Thus, in 1966 the total damage to agriculture in the United States was estimated at one-half billion dollars.[7, 8]

Two chemical species causing plant damage are particularly characteristic of photochemical smog. These are ozone (O_3)

and peroxyacetyl nitrate (PAN). Peroxyacetyl nitrate has the chemical formula

$$CH_3C \overset{\displaystyle O}{\underset{\displaystyle OONO_2}{\big\langle}}$$

PAN

and to anyone with a background in chemistry or explosives, it is obvious that this compound would be a highly unstable (explosive) and toxic material.[9] Indeed it is both, and PAN (and its higher homologs) is one of the major eye irritants, plant killers, and health problems in Los Angeles smog.

PAN has been found to be highly toxic to many important field crops, including romaine lettuce, alfalfa and spinach, and ornamental plants such as petunias, snapdragons and asters, to name a few.[6] Characteristically, with romaine lettuce the damage appears as a silvering on the *bottom* of the leaf.

In contrast, ozone, the other major phytotoxicant in photochemical smog, kills cells in the *top* surfaces of leaves and the damage appears as flecking or stippling of these upper surfaces. For example, a three-hour fumigation with air containing only 0.5 parts per million (ppm) of ozone produces lesions in grape leaves. In a recent review, Middleton notes that fifty-seven different species of plants are cited as being susceptible to ozone damage.[6]

Nitrogen dioxide (NO_2), a key toxic constituent of smog, about which we'll have more to say later, has an interesting effect on plants in that in fairly low concentrations it appears to restrict the growth of young plants without showing other common signs of injury.

Ethylene, the simplest olefin, $H_2C=CH_2$, is released by refineries and from auto exhausts and, specifically, is highly toxic to orchids in concentrations approaching the part per billion (ppb) range. The presence of this smog constituent has made it impossible for commercial growers to raise orchids in metropolitan areas of California.

Other air pollutants such as sulfur dioxide and hydrogen

fluoride are also serious phytotoxicants but, since they are not commonly associated with photochemical air pollution, I shall not discuss them here even though they are a problem in many areas of the United States.

We have now cited two general classes of pollutants present in photochemical smog: aerosols producing a dense haze and noxious gases such as ozone and PAN. Before going on, I want to stress the point that both the London and Los Angeles types of smog are *mixtures* of particulate matter and toxic gases. It is important to realize that in such a situation possible synergistic effects can occur which may exceed the sum of the effects due to the gases alone plus the effects due to the particles alone. In short, the whole may be greater than the sum of the parts. This is illustrated by the deadly effects of the London type of smog which contains both gaseous sulfur dioxide and solid particles of soot. The bulk of medical and epidemiological evidence to date strongly suggests that the 4,000 excess deaths in London in 1952 were due to synergistic respiratory effects arising from the simultaneous presence of *both* SO_2 and soot in the atmosphere. Similar considerations apply to the excess deaths reported in an acute smog episode in New York City in 1962. Air containing only SO_2 or only soot in concentrations close to those present in actual smog do not produce nearly as severe symptoms as does a mixture of the two pollutants.[10]

Obviously, one must be well aware of such synergistic effects when developing models for air pollution control. This point becomes clear when one examines material in two key documents on air quality criteria, one for oxides of sulfur and one on atmospheric particulate matter. They were published by the National Air Pollution Control Administration (NAPCA) in January 1969, and sent to the governor of each state who, by the terms of the Clean Air Act of 1967, will then have fifteen months to establish and implement air quality *standards* based on these criteria.[10]

A key question when it comes to modeling for control is just how valid are the data in these documents. Well, from intimate experience with these documents (we at ERI prepared preliminary drafts of both documents on a contract with NAPCA),

I can say that some data are quantitative (e.g., visibility reduction by particulates), some are qualitative and some are educated guesses (hunches). These are the kinds of present data which can and must be programmed into a computer model and we should all recognize this fact!

Accompanying the characteristic smoggy haze, due to photochemically produced aerosols, are the oxidizing gases, ozone, PAN, and nitrogen dioxide. Thus one can trace the buildup and movement of a smog cloud by use of strategically placed, continuously recording oxidant analyzers. Figure 2 (taken from Ref. 2) shows a typical change in oxidant concentration as a function of time of day for a single day (September 13, 1955) and averaged over three months (September-November 1955) at one sampling station in Pasadena. The maximum in oxidant

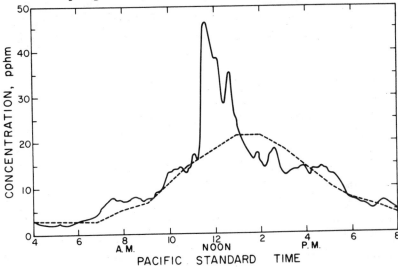

Figure   2.   Curves for oxidant as a function of time of day taken in Pasadena. The dash curve represents a three-month average of oxidant; the solid line the concentration on a single day. (From Ref. 2. Reprinted by permission.)

coincides with maximum solar intensity at noon. This can be contrasted to the graph of oxidant versus time of day taken with an oxidant analyzer on the UCR campus, Fig. 3. The

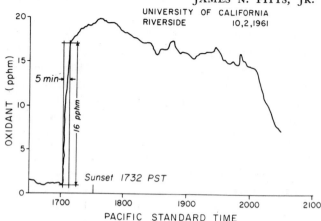
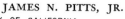

Figure   3.   Plot of oxidant concentration as a function of time
of day with an oxidant meter on the University of
California, Riverside campus, October 2, 1961.
Note the very low level of oxidant until 5:00 P.M.
and then the rapid rise as the smog cloud hits
Riverside.

oxidant was minimal at about 1 to 2 pphm until the cloud hit
and in a five-minute interval, starting at 5:00 P.M., the oxidant
jumped an increment of 16 pphm and *remained* at or above
the 15 pphm level until 8:00 P.M. This dramatic chemical
effect parallels the visible arrival of the smog cloud. Not only
can one see and instrumentally measure the arrival of the smog
on bad attacks, one can literally smell it, taste it, and be made
ill by it! If a person is strenuously exercising, the smog
literally hits him, cutting down his wind and vitality.

Along with the haze and gas cloud comes a somewhat more
subtle and often overlooked effect, the reduction of solar
energy, particularly in the ultraviolet, reaching the earth's
surface. This is dramatically shown in Fig. 4 (from Ref. 2)
which shows the calculated solar radiant energy in a 10 Å
band centered at 3235 Å on a clear day, incident on a normal
surface (solid line), and the actual energy observed on a smoggy
day by Stair in Pasadena, October 18, 1954 (dotted line).

Let us now go into more detail on several aspects of the
chemistry of urban atmospheres, in particular, the chemistry
of photochemical type smog.

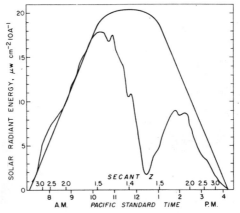

Figure 4.   Plot of solar radiant energy striking the earth's surface at Pasadena on a given day. The curve with the smooth, single maximum shows the amount of energy calculated to fall in a 10 Å band centered approximately 3250 Å. The irregular line with the deep minimum is the actual observed light intensity as determined by Stair in Pasadena. (From Ref. 2. Reprinted by permission.)

There are two broad types of pollutants that we must deal with when we develop mathematical models, *primary pollutants* which are those emitted directly to the atmosphere and *secondary pollutants* which are formed by chemical or photochemical reactions of the primary pollutants, after the latter have been emitted into the atmosphere and have been exposed to the sunlight.

Figure 5, provided by the Los Angeles County Air Pollution Control District, was compiled in January 1967 and gives average values in the *tons per day* and the percentage contribution of the important contaminants from the major sources within Los Angeles County. All of the compounds cited are primary pollutants, that is, the organic gases which we will refer to in the future as hydrocarbons, the oxides of nitrogen (NO_x), sulfur dioxide (SO_2), and carbon monoxide.* It is stag-

---

* Recall that NO_x actually stands for a mixture of NO and NO_2 in an unspecified ratio. As it emerges from the auto exhaust, the NO_x is almost entirely nitric oxide, NO. This is important to recognize when developing atmospheric models.

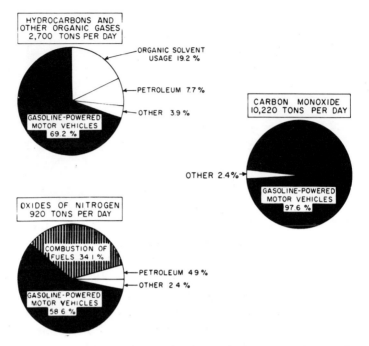

Figure   5.   Percentage contributions of air contaminants from
                major sources in Los Angeles County. Data pre-
                sented by Los Angeles County Air Pollution Con-
                trol District, January 1967.

gering to realize that almost *15,000 tons per day* of pollutants
are introduced into the atmosphere over Los Angeles County.
On the other hand, it is even more staggering to consider what
might have been the case if the Air Pollution Control District
of Los Angeles County had not been taking stringent measures
to prevent the emission of pollutants from stationary sources.
For example, the current program of pollution control prevents
the emission of approximately 5,000 tons of pollutants daily in
Los Angeles County.

   In terms of the per cent contribution to air pollution from
various major sources, note that the gasoline powered motor
vehicles in Los Angeles County account for approximately
seventy per cent of the total hydrocarbons, sixty per cent of
the oxides of nitrogen, six per cent of sulfur dioxide, and about
ninety-seven per cent of the carbon monoxide. Thus, gasoline

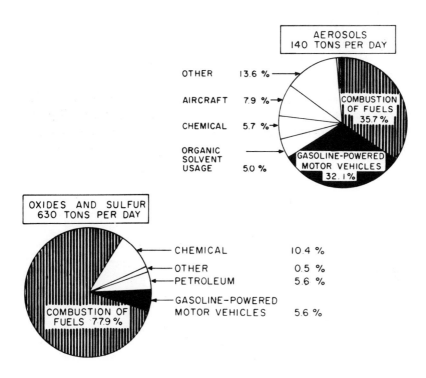

Figure   5. Continued

powered motor vehicles account for approximately eighty-five per cent of the total air pollution. Also significant is the fact that thirty-two per cent of the aerosols released into the atmosphere come from automobiles and trucks.

When these *primary pollutants* emerge from the auto exhaust, they are diluted with approximately 1,000 volumes of air per volume of exhaust gas producing actual atmospheric concentrations of the order of 0.5 ppm NO_x, 1 ppm hydrocarbons and 5-20 ppm CO. Actually, these are order of magnitude average concentrations; actual cases in the Los Angeles Basin may be much higher. For example, maximum air contaminant concentrations in the Los Angeles Basin in 1967, as reported by the Air Pollution Control District, County of Los Angeles, are carbon monoxide, 72 ppm; hydrocarbons, 40 ppm

232   JAMES N. PITTS, JR.

(flame ionization method); oxides of nitrogen, 3.9 ppm; ozone,
0.9 ppm; and sulfur dioxide, 2.5 ppm.

When an urban atmosphere of this type is irradiated by the
sun, a series of highly complicated photochemical reactions
occur. Products of these photochemical reactions are the *secondary pollutants* referred to earlier. They are responsible for
eye irritation and plant damage and include nitrogen dioxide,
ozone, PAN, formaldehyde, higher saturated aldehydes such as
acetaldehyde, unsaturated aldehydes such as acrolein (a very

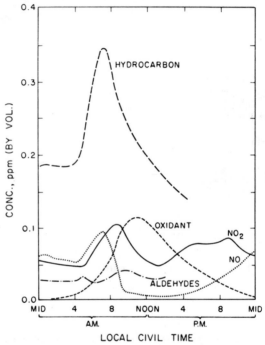

Figure 6. Plot of average concentrations of hydrocarbons
oxidant, nitrogen dioxide, nitric oxide, and aldehydes in the atmosphere as a function of the time
of day in downtown Los Angeles. The data were
generated by the Los Angeles County Air Pollution Control District for days when eye irritation
was present. Data for hydrocarbons, aldehydes, and
ozone (1953-1954); NO and NO$_2$ (1958). (From
Ref. 2. Reprinted by permission.)

powerful eye irritant), alkyl nitrates, alkyl nitrites, and a variety of other compounds.

Again for modeling purposes, we should note that the concentrations of these secondary pollutants vary in a complex way with atmospheric conditions, time of the day, etc. This is shown in Fig. 6. In following these variations, one is interested not only in an instantaneous concentration or in a total dose, both of which are important, but also the concentrations of the important pollutants as a *function of time*. Such instruments, which provide this information, are now available and, although not totally satisfactory in all cases, they provide data giving a reasonable picture of the diurnal variations of the concentrations of major pollutants. These data are key items when one develops a diffusion model for reactive pollutants.

Carbon monoxide is a particular interest, not only because of its health effects but also in that it is quite unreactive in the atmosphere. Thus it serves as a useful tracer to indicate the activities of automobiles because, for example, as seen in Fig. 5, almost ninety-eight per cent of the carbon monoxide released into Los Angeles atmosphere is from gasoline powered motor vehicles (*on the average, twenty-nine pounds of CO are emitted per ten gallons of gasoline consumed*). A typical plot of the concentration of carbon monoxide in ppm as a function of time in the Los Angeles-Pasadena area shows the concentration dropping to a minimum at approximately 5:00 A.M., then rising rapidly to a maximum about 8:00 A.M. This makes sense when one recognizes that this is the time of peak activity on the freeways, and also at this time there is generally a stable atmospheric inversion which traps the carbon monoxide. A much broader and lower peak occurs between 4:00 and 8:00 P.M. when the cars again hit the freeway. Here the effect is not as noticeable because the atmospheric inversion is not as intense and the carbon monoxide is dispersed by the breezes.

Because of its nonreactivity, the fact that it comes largely from automobiles, and because excellent analytic instrumentation is available, CO was chosen as the key pollutant to be considered first in developing the preliminary model requested in the CRC-APRAC proposal mentioned earlier.

Now, as an example of modeling, I should like to consider briefly the approach of the MRI-SDC-ERI team to the previously mentioned RFP for "Developing and Validating Mathematical Models and Methodologies for Estimating Air Pollutant Concentrations at Various Locations in Any Urban Area." [11] The first phase of our approach placed strong emphasis on the formulation of sound physical-statistical mathematical models for each of the subsystem components of a total system model. Among these are driving activity, vehicle emission characteristics, traffic flow, stationary sources, concen-

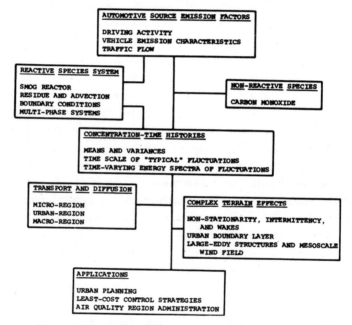

Figure 7.   Interrelationships between system submodel components for a general model. (From Ref. 10.)

tration-time histories, turbulent diffusion, transport, urban boundary layer complexities, depletion processes, and atmospheric photochemical processes. The interaction between such submodel components to produce a *general model* for pollution from automotive sources is shown in Fig. 7.[12]

A key problem in the early stages of development of a general model is the acquisition and judicious selection of already

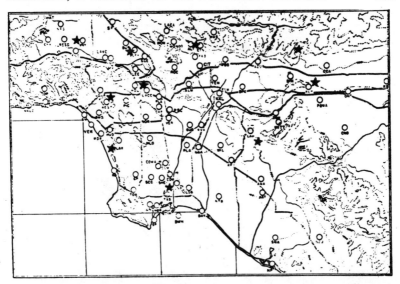

Figure   8.   Main automobile transportation network, CAMP-
like (*) monitoring stations and meteorological
data stations for the Los Angeles Basin. (0).
(From Ref. 10.)

existing data for several different types of urban areas. Actually,
reliable and extensive data bases for the construction and
validation of sophisticated *general models* do not yet exist for
any urban area. One of the most extensive data collecting
networks in the United States exists in Los Angeles County
(Fig. 8). However, considering the vast area of the County, it
is clear that more air monitoring stations would be most useful.
On the other hand, Los Angeles is fortunate indeed compared
to most major U.S. cities which have fewer such facilities.

Data from the Los Angeles grid of monitoring and meteorological stations were, incidentally, utilized by my co-workers
Mr. Schuck, Dr. Wan, and myself in a paper which represented
one of the early, and necessarily unsophisticated, efforts to
develop a preliminary version of a type of submodel for
photochemical pollution in an urban area.[13]

CAMP stations are set up in a number of United States
cities; they provide a highly useful source of data on atmospheric concentrations of key pollutants. For the key air pollu-

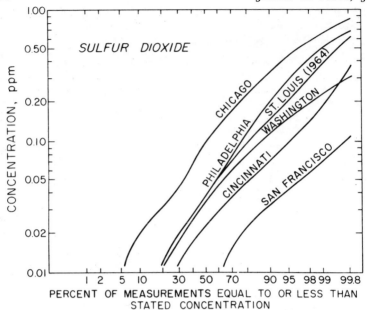

Figure 9. Frequency distributions of atmospheric sulfur
dioxide levels of six U. S. Cities (one hour averaging time; original source, CAMP.) (From Ref. 12.
Reprinted by permission.)

tant sulfur dioxide, Fig. 9 is an example of the type of information useful for atmospheric models that can be derived from
data collected in this program.[10a]

Also critical to the development of the submodels for pollution from automotive sources is an accurate and detailed assess-
ment of the utilization of the various portions of the transportation network distributed throughout a given urban area. Traffic
flow data are useful, but for precise micro-modeling the urban
area, one needs to know the traffic density on an hourly basis
distributed over the entire twenty-four hour diurnal cycle, the
weekly cycle expressed on a daily basis, seasonal traffic flows,
and the traffic pattern for special cases such as sporting events.
For example, in the Los Angeles Basin we found that the concentration of maximum daily oxidant is a function of the day
of the week and that this function might be related to automotive traffic patterns.[13] Thus, for certain time periods, maximum daily oxidant was at a *minimum* on weekends at stations

in the northern portion of the Basin (Burbank, Pasadena, Azusa; see map, Fig. 1), but at a *maximum* near Inglewood and Long Beach, cities in the southern portion of the Basin. These two distinctly different weekend patterns correspond in part to the recreational activities of the populace and, in addition, indicate that mixing is less than complete in the Los Angeles Basin.

I should like to conclude by briefly mentioning some aspects of research on air pollution chemistry currently being carried out in our laboratory at UCR. Some of this work is relevant to the question of a model for a photochemical reactive air pollution system.

It is well-known that if one starts out in the laboratory with a reaction chamber containing only nitric oxide, a trace of nitrogen dioxide, and air, and irradiates this mixture with ultraviolet light, the following reactions occur.[2-4, 9b]

$$NO_2 + h\nu \rightarrow NO + O$$
$$O + O_2 + M \rightarrow O_3 + M$$
$$O_3 + NO \rightarrow O_2 + NO_2$$

The net effect of irradiation on this inorganic system is to set up a dynamic equilibrium, namely

$$NO_2 + O_2 \underset{\leftarrow}{\overset{uv}{\rightarrow}} NO + O_3.$$

However, if a hydrocarbon, particularly an olefin or an alkylated benzene, both of which are common constituents of gasoline, is added, the dynamic equilibrium is unbalanced and the following events take place:

1. The hydrocarbons are oxidized and disappear.
2. Reaction products such as aldehydes, nitrates, PAN, etc. are formed.
3. Nitric oxide is converted into nitrogen dioxide.

When all of the nitric oxide has been used up, ozone starts to appear. On the other hand, PAN and the aldehydes are formed from the beginning of the reaction.

One can reproduce the essential features of photochemical air pollution by simply irradiating dilute automobile exhaust in a suitable reaction chamber. The auto exhaust disappears

238 JAMES N. PITTS, JR.

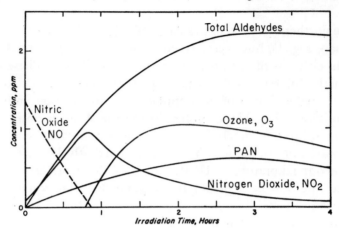

Figure 10. Variation of concentration of various components
in auto exhaust irradiated with ultraviolet light.
Data of this type have been reported by Schuck
and co-workers.

and various compounds are formed identical to those found
in actual urban atmospheres. Figure 10 shows the variation
with time of some typical products of ultraviolet irradiation of
auto exhaust in a laboratory reaction chamber.[2] Note that the
chemical behavior in this system, as a function of time, resembles that seen in Fig. 6, which gives the actual diurnal variations
of pollutants in the atmosphere.

The curves shown in Fig. 10 were generated from data taken
in a rather unique reaction system, the key feature of which is
a long path infrared cell. We are currently using such a system
at UCR. A diagram of the reaction cell is shown in Fig. 11.
The basic idea of this apparatus is that one employs a conventional infrared spectrophotometer for analysis of the gas but
instead of using small gas cells of about 0.1 liter capacity, one
employs two large cylindrical tanks, each of approximately
sixty liter volume. One acts as a blank cell and the other as a
photochemical reaction chamber.

Basically, two beams emerge from the infrared source. One
traverses the blank cell, which is filled with the nonirradiated
material, and then goes into the infrared detector. The other
beam goes into the reactor cell. This is fitted with quartz

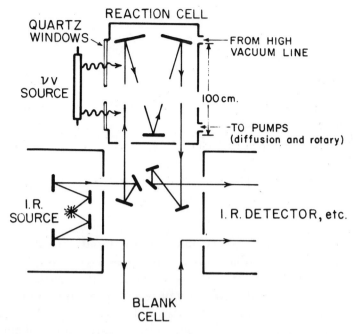

Figure 11.   Diagram of optical path for long path length infrared spectroscopy.

windows so that the reaction mixture in the cell can be irradiated with an external ultraviolet light source. The mirrors in the reaction cell and the dummy cell can be aligned so that the infrared analyzer beam traverses a path of up to forty meters before leaving the reaction cell. Compared to the 0.1 meter path in a conventional gas cell, this provides a tremendous sensitivity and, with this instrument, one can quantitatively identify and determine various pollutants such as olefins, ozone, etc., in the fraction of part per million range of concentration.

A conclusion important for modeling purposes arises from experiments in long path length infrared cells of this type by such researchers as Stephens, Schuck, and Tuesday (see Refs. 2, 3, and 4). The various hydrocarbons emitted in auto exhaust have *a wide range of reactivities in forming photochemical smog*. That is, hydrocarbon A introduced into a mixture of nitric oxide and air and irradiated might react very much more

Table 2

Relative Hydrocarbon "Photochemical Reactivities"
in the System HC-NO-NO₂-AIR *

Substances	Molar Reactivity
C_1-C_3 Paraffins	0
C_4 and Higher Paraffins	1
Ethylene	4
1-Alkenes	7
Internally Double Bonded Olefins	8
Acetylenes	0
Benzene	0
Toluene and Other Mono-Alkyl Benzenes	3
Dialkyl and Trialkyl Benzenes	6

* Data of John Maga, California Department of Public Health, September 1966.

slowly or, conversely, very much more rapidly than hydro-carbon B under similar conditions. This is not unexpected since the different hydrocarbons have different chemistries and one would expect different rates of reactions for different types of hydrocarbons. Various hydrocarbon reactivity indexes for use in air pollution control have been prepared. One of these, developed by Mr. John Maga, Executive Secretary of the Air Resources Board of California, is shown in Table 2.

It is also significant that Professors Friedlander and Seinfeld, at the California Institute of Technology and MRI, are devel-

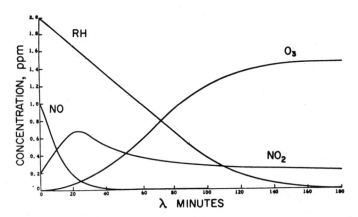

Figure 12.   Computer time-concentration curves for model of photochemical air pollution. (From Ref. 10.)

oping a useful kinetic-mechanistic model which, when fed into a computer, produces concentration-time curves (Fig. 12) in good agreement with many aspects of the actual curves for irradiated auto exhaust shown in Fig. 10.[10]

One further point should be noted. When we consider models for urban atmospheres, we should clearly recognize that we do not really know in detail the distribution and concentrations of *all* contributors to photochemical smog. This is particularly the case for highly reactive transient species. Thus, recently our group at UCR has proposed that a new and possibly important species, electronically excited singlet molecular oxygen, symbolized as $O_2(^1\Delta)$ [plus possibly $O_2(^1\Sigma)$], may be involved in a variety of systems, including test chambers, as well as in polluted atmospheres. We furthermore believe that hitherto unexplained chemical and possibly biological effects on man, animals, and plants might be attributed to these species. In June of 1967, we presented a preliminary paper in which we suggested that singlet molecular oxygen, which is normal molecular oxygen, $O_2(^3\Sigma)$, raised to its first electronically excited state by the addition of approximately twenty-two kilocalories of energy, might be formed by an energy transfer process.[14] In this process, organic molecules in the atmosphere absorb solar radiation and then on collision with normal oxygen molecules transfer their electronic energy to the oxygen molecules producing highly reactive singlet oxygen, $^1O_2$. Of particular importance is the fact that electronically excited singlet oxygen reacts with olefinic substances to produce thermally unstable hydroperoxides. Such hydroperoxides clearly could be involved in a variety of chemical and biochemical processes.*

In addition to energy transfer, we have proposed several other possible mechanisms for the formation of singlet molecular oxygen in the atmosphere.[15][16] Direct photolysis of an atmospheric contaminant could form singlet oxygen in the primary step. Thus, in the laboratory, the ultraviolet photolysis of ozone is known to produce singlet oxygen

---

* For detailed reviews and original references to the properties and reactions of singlet oxygen, see Foote,[17] Gollnick,[18] and Wayne[19] and for environmental considerations, see Pitts, *et al.*[15][16]

$$O_3 + h\nu \ (\lambda < 3200A) \rightarrow {}^1O_2 + O$$

and ozone is not only present on a world-wide basis in the upper atmosphere, but also in smoggy urban atmospheres. Other possible sources of singlet oxygen are exothermic chemical reactions involving atmospheric contaminants. We are currently in the process of examining a variety of these interesting possibilities.

In conclusion, I would like to acknowledge my indebtedness to my research group and to the National Air Pollution Control Administration which has generously supported our research through Grants AP 00109 and AP 00771. I also wish to acknowledge a sabbatical leave from the University of California during which time the manuscript was prepared.

REFERENCES

1. Jack McKee, Letter to the Editor, *Los Angeles Times,* February 8, 1968.
2. P. A. Leighton, *Photochemistry of Air Pollution* (New York: Academic Press, 1961).
3. A. C. Stern, ed., Air Pollution, 2nd ed., vols. I, II and III (New York: Academic Press, 1968).
4. A. P. Altshuller and J. J. Bufalini, "Photochemical Aspects of Air Pollution: A Review." *J. Photochem, Photobiol., 4,* 97, 1965.
5. A. J. Haagen-Smith, E. F. Darley, M. Zaitlin, H. Hull, and W. Noble, "Investigation on Injury to Plants from Air Pollution in the Los Angeles Area." *Plant Physiol., 27,* 18, 1952.
6. J. T. Middleton, "Air Pollution Threat to Flora and Fauna." *Conservation Catalyst, 2,* 1967.
7. U.S. Department of Health, Education and Welfare, *The Effects of Air Pollution.* U.S.P.H.S. Publication 1556, 1966.
8. For an excellent review of Air Pollution Damage to Plants see the Symposium papers published in *Phytopathology, 58,* 1075, 1968.
9. (a) E. R. Stephens, "The Formation, Reaction, and Properties of Peroxyacyl Nitrates in Photochemical Air Pollution." in *Advances in Environmental Sciences,* vol. 1, J. N. Pitts, Jr. and R. L. Metcalf, eds. (New York: Interscience Publishers, 1969), p. 119; (b) E. Schuck and E. R. Stephens, "Oxides of Nitrogen." *ibid.,* p. 73.
10. (a) "Air Quality Criteria for Sulfur Oxides." U.S. Department of Health, Education and Welfare, Public Health Service, Consumer Protection and Environmental Health Service, National Air Pollution Control Administration, Washington, D.C., January 1969, NAPCA Publication AP-50; (b) "Air Quality Criteria for Particulate Matter." *ibid.,* January 1969, NAPCA Publication AP-49.

11. "Developing and Validating Mathematical Models and Methodologies for Estimating Air Pollutant Concentrations at Various Locations in any Urban Area." Main Technical Proposal, prepared by Meteorology Research, Inc., Altadena, Calif., with the active involvement of Environmental Resources, Inc., Riverside, Calif., and Systems Development Corp., Santa Monica, Calif., September 1968.

12. D. H. Slade, "Modeling Air Pollution in the Washington, D.C. to Boston Megalopolis." *Science, 157,* 1305 (1967).

13. E. A. Schuck, J. N. Pitts, Jr. and J. K. S. Wan, "Relationships Between Certain Meteorological Factors and Photochemical Smog." *Int. J. Air Water Poll., 10,* 689 (1966).

14. A. U. Khan, J. N. Pitts, Jr. and E. B. Smith, "Singlet Oxygen in the Environmental Sciences; the Role of Singlet Molecular Oxygen in the Production of Photochemical Air Pollution." *J. Env. Sci. and Tech., 1,* 656 (1967).

15. J. N. Pitts, Jr., A. U. Khan, E. B. Smith and R. P. Wayne, *ibid., 3,* 241 (1969).

16. J. N. Pitts, Jr., "Photochemical Air Pollution: Singlet Molecular Oxygen as an Environmental Oxidant." in *Advances in Environmental Sciences,* vol. 1, J. N. Pitts, Jr. and R. L. Metcalf, eds. (New York: Interscience Publishers, 1969).

17. (a) C. S. Foote, *Accounts of Chemical Research, 1,* 104 (1968); (b) *Science, 162,* 963 (1968).

18. K. Gollnick, "Type II Photo-Oxidation Reactions in Solution." in *Advances in Photochemistry,* vol. 6, W. A. Noyes, Jr., G. S. Hammond and J. N. Pitts, Jr., eds. (New York: Interscience Publishers, 1968), p. 1.

19. R. P. Wayne, "Singlet Molecular Oxygen." in *Advances in Photochemistry,* vol. 7, J. N. Pitts, Jr., G. S. Hammond and W. A. Noyes, Jr., eds. (New York: Interscience Publishers, 1969), p. 400.

Weather Modification Research—A Desire and an Approach

S. M. GREENFIELD*

The RAND Corporation
Santa Monica, California

The continuous and natural involvement of man and his environment has led him to a constant desire to influence or control it to his benefit. Most common of the attempts in this direction, and certainly the oldest, is the hallowed appeal for more rain, or better weather, or a calm sea, etc., contained in every religion known to man. At the very least, such a religious approach indicates that man, through the ages, has been extremely interested in, and willing to think about, the possibility of not being subjected to the inevitability of the weather.

In a more modern vein, the growth of science and technology, in the absence of good communication with the lay public, very often spawns a pseudo-science which steps in to fill this gap. Such was the case in the United States during the nineteenth century, during which period the itinerant rainmaker practiced his trade on a public that was all too ready to

* Any views expressed in this paper are those of the author. They should not be interpreted as reflecting the views of the RAND Corporation or the official opinion or policy of any of its governmental or private research sponsors.

believe the abilities of science to solve this environmental problem.

This strong desire of man to control his environment has persisted, and with the growth of understanding of atmospheric dynamics and physics has blossomed into a more directed scientific interest in weather modification and control. Although some work had been done in cloud physics and rainmaking prior to World War II, the real birth of this field occurred in about 1946 with the work of Langmuir and his colleagues of The General Electric Research Laboratories.[1] Langmuir demonstrated in the laboratory that supercooled water droplets could be nucleated with dry ice crystals, thereby freezing and precipitating. He repeated this experiment in a supercooled cloud in the atmosphere and without a doubt modified the cloud. It should be pointed out that, while the photographic evidence showed a clear hole carved in the cloud at the point of "seeding," no claim for providing rain was made at that time. The rainmaking claim for cloud seeding, however, quickly followed.

This experiment was followed by several additional attempts by the same group to utilize the technique to produce a controlled weather effect. In addition to precipitating ice crystals, these attempts inadvertently precipitated an almost frantic activity in rainmaking in at least one segment of the meteorological community that persisted for almost two decades with some interesting results.

First, it created a great deal of interest in the lay public in an activity called rainmaking and cloud seeding attaching to it, somewhat prematurely, a little understood, scientific facade. Second, this trend towards commercialism and the public claims of success had an even more important effect, in that they kept the majority of good researchers out of this field for almost twenty years. We are only just beginning to recover from that period. It should be recognized that this effect was only indirectly the result of Langmuir's research. He was too good a researcher himself to have desired the immediate attempts to exploit untested ideas. Rather what happened was that the public was anxious to accept this as a feasible idea, and demanded that it be utilized. Such demands will always attract those who are willing to react regardless of the predictability of the outcome.

The initial impetus to weather modification as a scientific subject was provided in the late 1950's when the National Science Foundation (NSF) was given a charter by the Congress to "initiate and support a program of study, research, and evaluation in the field of weather modification." * In this general atmosphere of reluctant researchers, NSF attempted to set up a research program primarily in the universities. Unfortunately, attracting a broad spectrum of researchers to this field continued to be very difficult and for many years the major portion of these funds went to support cloud seeding or cloud physics types of experiments in the atmosphere. What was lacking in this program was the very necessary strong theoretical and laboratory efforts which are particularly important because in this medium it is next to impossible to conduct a controlled experiment on any meaningful scale. Under these conditions, the theoretical and laboratory portions of the program, properly interacting with the experimental, provide the most probable means of achieving the desired understanding of the phenomena under investigation.

It would be incorrect to say that the NSF program was completely unbalanced. Some laboratory work was stimulated and in several universities there was a growing interest in looking at basic problems such as the coalescence of droplets, the condensation process, the ice nucleation process, and the various other parameters that bear on whether or not seeding produces rain.

The large impetus to weather modification research as a broad field actually occurred in the early '60's. During this period two research documents appeared which attempted to examine the possibilities of weather modification research.[2, 3] Both pointed to major revolutions in the atmospheric sciences that had occurred or were on the immediate horizon; the first

---

* Public Law 85-510.

to the growing power of large-size computers and the apparent success that was attendant on attempts to numerically model the atmosphere at the storm or synoptic scale. This success was already manifest in the operational short-range numerical forecasting effort. Further, there was the deep active commitment by several competent research groups to modeling the atmosphere at the global scale using the so-called nonlinear primitive equations of motion. This latter approach offers the possibility of numerically modeling the entire global atmosphere.

The second important revolution was the success of the so-called space age and the emergence of the growing capability to observe atmospheric parameters all over the globe essentially simultaneously. This latter point offered the real possibility of being able to acquire the global data required for the large numerical models.

In addition, the point was made that there was some very interesting laboratory and theoretical work that should be started at the cloud physic end of the atmospheric scale involving the question of how cloud droplets achieve precipitation sizes.

Finally, both of these studies pointed out the possible danger of attempts to modify large-scale weather phenomena without sufficient understanding of the possible results. The danger here, of course, lies in the fact that such experiments could lead to irreversible changes in the environment, deleterious to man. One example that comes to mind has seriously been suggested on several occasions. This is the suggestion that removing the Arctic sea ice would radically improve the climate of Siberia. The danger lies in the fact that, in addition to improving the Siberian climate, at least one logical chain of argument suggests that the ice would not return for several thousands of years and that its absence would precipitate a new Ice Age.[4] The disturbing element in this example is that we cannot at this time decide whether this logical chain is correct. Yet at the same time, due to the growth of man's energy producing capabilities, and the tenuous stability of the Arctic sea ice, there is a distinct possibility that this event could occur either deliberately or inadvertently. Further, this might occur prior to our determination of the results of such an event and of the possible courses of action to either avert or reverse the trend. The large computers and the global atmospheric models, given the proper support, offer the necessary tools that might permit us to investigate these possibilities.

Having discussed the necessity of understanding the atmospheric environment in depth as a prime requisite for weather modification, it is necessary to examine, at least diagrammatically, the nature of the problem. Figure 1 is an attempt to provide such a diagram. Let us start by considering the atmospheric scales of motion as representing a spectrum on which one can impress the various observed phenomena. It is found that such a size spectrum extends over essentially sixteen or seventeen orders of magnitude from the molecular to that of the planetary circumference. This is the scale shown at the top of this figure. As is evident from this illustration, understanding the atmosphere involves all of the familiar phenomena that one normally thinks of as weather. In addition, however, it also involves understanding the exchange or input of energy from external sources or sinks (i.e., the direct or indirect input of solar energy, the radiation of energy to space, and the interaction with the underlying surface), as well as the internal interaction between scales. Within the figure, the solid arrows represent interactions that we know occur, solid lines the known scale effect of these phenomena, and the dotted arrows and dotted lines are used to represent a degree of uncertainty. We know, for example, that a large fraction of the energy input to the atmosphere comes through the evaporation-condensation mechanism, whereby water is evaporated from the surface, lifted into the atmosphere, and, in the process of recondensing, releases the latent heat to the atmosphere. This occurs at approximately the cloud scale of motion as indicated. We know that the large eddies, representing the general circulation and driven by the pole-equator temperature gradient, ultimately break down into smaller and smaller eddies and finally dissipate as heat at the molecular scale. This process is shown on Fig. 1 as the large open dashed arrows indicating the transfer of energy down-scale to the point of dissipation.

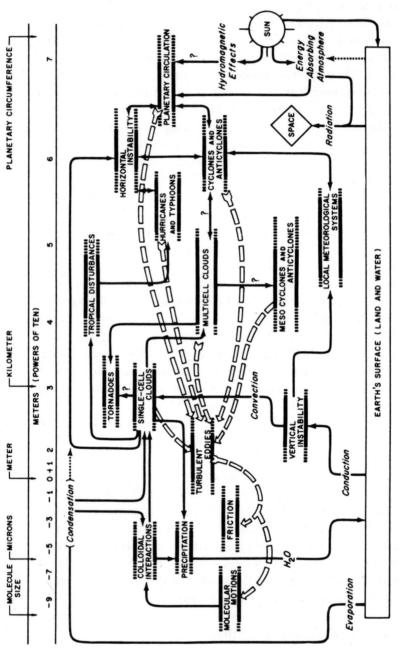

Figure 1.   Atmospheric interactions.

Although it is very difficult to prove, we suspect that the energy must also be transferred up the scale of atmospheric motion. The reason for this lies in the fact, previously mentioned, that a large fraction of the energy input to the atmosphere occurs in the cloud-size, vertical-instability portion of the spectrum. The energy at this scale must contribute, in part, to the horizontal instability that results in the larger weather phenomena and hence, probably through an organizational process involving feedback, moves up-scale.

It is clear that an initial, and possibly primary, part of this problem involves an understanding of the energetics of the atmosphere. A preliminary attack on the problem might be to attempt to determine where the major instabilities exist on the scale of atmospheric motions, and how the energy distributes itself in the system.

Figure 2 is a very elementary attempt to arrive at this preliminary answer. It is based on the fact that a time history of wind velocity recorded at a point can be expressed by a Fourier analysis of the time variations. From the relationship of kinetic energy to the square of the velocity, each Fourier component can be expressed as the square of the Fourier coefficient. Hence one can interpret the velocity-time record in terms of the energy available as a function of frequency. By assuming some average velocity one may convert the frequency to a scale of motion sizes, as in Fig. 2. It should be noted that the energy scale is arbitrary, hence the curve in Fig. 2 should only be considered, at best, a description of the relative energy distribution (at one hopefully typical point) and should only be an indication of what ranges of motion size contain the bulk of the available kinetic energy. Despite these limitations it is interesting to note that, as one might expect, the two energy maxima are located at the cloud-size, representing the vertical-instability, and the synoptic-size, representing the major horizontal-instability. As is well known, these two forms of instability represent the basic driving forces in the atmosphere, and hence probably provide the most fruitful avenues of approach to possible weather modification.

Unfortunately, the atmosphere is not so simply described

either energetically or dynamically. For example, the region of low available kinetic energy between the two peaks is called the mesoscale of motion, and represents a little understood class of phenomena. Does it really contain little average energy because it represents a transitory region, or one containing little activity? Or is this apparent feature due to the fact that the point of measurement is atypical of this class of motion? Such uncertainty in as basic an atmospheric parameter as energy distribution must give one pause when considering the advisability of conducting extensive atmospheric experiments without acquiring the necessary understanding.

The question then is how to proceed in this field. You would like to have the ability to conduct atmospheric experiments. You are faced, however, with a reluctance to experiment at the larger scale in the absence of understanding. At the smaller scales, where one can presumably experiment safely, there exists the frustrating inability either to control the experimental condition or measure the experimental results adequately. In this case, as in the case of the larger scales, there is a great attraction to turn to numerical modeling as an aid to understanding. With such models available, then the question of how to use them most expeditiously represents a quandary currently facing the atmospheric sciences, particularly with regard to weather modification. It has been assumed in the past that the numerical models could be the end in themselves, that is to say it has been implied that, with enough time and computer power, adequate atmospheric models would result, and then weather modification research would proceed very quickly. However, this now appears to be a simplistic approach and that the apparent need for numerical models is one calling for a growing interplay between numerical, laboratory, and field experimentation.

In a recent RAND research report [5] three basic categories of application of modeling to weather modification research were identified; these are contained in the following quotes.

> First, numerical modeling may be applied in conjunction with field or laboratory experiments. This application offers a potentially powerful technique for advancing our knowledge of atmospheric processes,

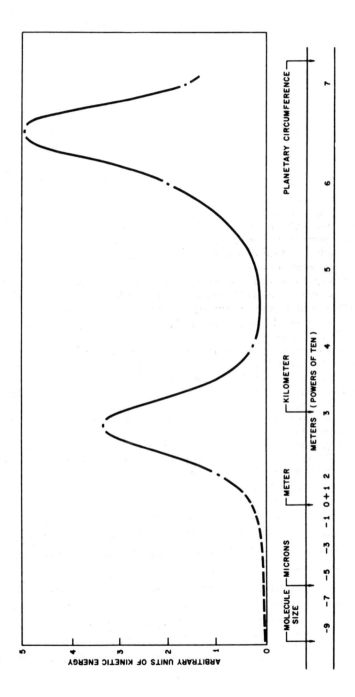

Figure 2.   Atmospheric available kinetic energy distributions.

particularly at the small end of the scale of motion. In this case one recognizes the uncertainties that currently exist both in our own ability to explain what we observe in the atmosphere and in our estimate of the adequacy and accuracy of our models. One can visualize a very important place for the numerical model in helping in the *design* and *evaluation* of a stratified set of field experiments directed toward removing a degree of uncertainty. The resulting improved understanding of the physical processes could lead to an improvement in the model, and hence to further stratified experimentation. Such an iterative feedback might ultimately lead both to a much improved knowledge of atmospheric phenomena and to numerical models capable of reliable simulation.

Second, given the ability to simulate the atmosphere at a specific scale, one conceivably could utilize the numerical model to aid in the evaluation of field experiments. In this case one can visualize models occupying a spectrum of capabilities classified by the accuracy and detail available in the model. The more accurate predictive model could provide the experimental control normally missing in any meteorological field experiment. The less accurate model could provide a categorization of the atmospheric conditions under which the experiment was run. In essence, this would be an after-the-fact stratification of the experiment providing, where applicable, a considerable improvement in the significance of the statistical evaluations.

Third, given the ability to simulate the atmosphere reliably at a specific scale, one might then utilize a numerical model as a mechanism for exploring the possible results of suggested weather-modification attempts. In this case one is truly using the model to perform a 'numerical experiment.' The confidence in the 'experimental' results certainly depends on the accuracy with which the model calculates the atmospheric processes at the scale of interest. However, it is not at all clear that less detailed and hence less predictive models do not possess some degree of ability to test proposed weather modifications. In essence, one suspects that a model that is internally consistent and physically valid (albeit coarse) still might indicate the general trend that will be produced by a specific modification technique on a specific atmospheric situation.

When one talks, for example, of possible modification at the large end of the scale of weather phenomenon, the planetary circulation scale, one is talking about effects that might build up over tens or even hundreds of years of real-time. The thought of running a model for a hundred years of real-time, regardless of what this means in machine time, is very troublesome, particularly if one must worry about providing sufficient detail to insure a perfect reproduction of the atmosphere. The third category suggests that it is probably not necessary to achieve such perfection in order to significantly examine weather modification possibilities at least at the larger scales of motion. What may be possible is that, if one is sure that there are no internal inconsistencies in the model nor significant gains or losses of energy or mass in the system, then one might use this as a difference model. That is, one can make two parallel runs in time, one involving the perturbation, and one not, and examine the results. Such an experiment may not provide a detailed prediction of the results, but in view of the above discussion it certainly may be able to provide significant insight into the general response of the atmosphere to such an event.

In summary then I can quote once again from the above RAND report:

> It is apparent, therefore, that numerical models offer a wide spectrum of ability in any program of research on weather modification. Furthermore, it is also apparent that one need not wait for the model possessing the 'ultimate accuracy' to obtain some degree of practical utility. What is obvious is that the *degree* of accuracy *must* be known, and that for many uses, the full potential of the numerical model cannot be realized when employed isolated from the laboratory/field experiments. If weather-modification research is to proceed at all expeditiously, the numerical modelers and the theoreticians must be brought into close cooperative contact with the laboratory and field experimenters. This area of research cannot permit the condition to exist where the numerical models become ends in themselves and researchers become mesmerized by their own abilities to utilize high-speed computers to 'simulate' the real world and hence lose contact with the very realities they hope to duplicate. Nor can we permit the condition of endless experimentation gathering masses of data with no hope of disentangling the confusion produced by the accustomed lack of experimental control. To date, both of these conditions exist to a depressingly wide extent.

The First International Weather Modification Symposium was recently held at the State University of New York at Albany. This symposium attempted to bring together the numerical modelers and the experimenters, as well as people working at various points on the scale of atmospheric motion. It was clear from this gathering that there is little communication outside of the immediate areas of personal interest. There are very few examples of experimenters in the field communicating with the modeler at their scale of interest, and seeking aid in experimental design and evaluation. Nor was this apparent lack of communications one sided. The numerical modelers, in general, appeared to act as if the real world were captured on their computers. Such a situation is guaranteed to keep good researchers working blindly in their own small region of the spectrum never looking up to interact with the problems occurring at the surrounding scales of motion. It also guarantees that we will never achieve weather modification (perhaps weather control is the better phrase; weather modification we may achieve or stumble into and then wish we had not), because this approach will never lead to a full understanding of the total atmosphere.

In addition to just the natural reluctance of researchers to waste time communicating out of their areas of immediate interest, there may be two other reasons for this tendency towards noninteractive research. One of these reasons is certainly contained in the manner in which we are currently funding weather modification research in this country. Figure 3 is a graph from the RAND report, illustrating the research funding problem. It shows the distribution of federal government funding in cumulative amounts, in this area, for two years, 1962 and 1966. In 1962, there were twenty-two field experiment programs totaling 1.86 million dollars. In addition, there were fifty-five research projects concerned with the theoretical investigation of atmospheric phenomena totaling 2.71 million dollars. By the end of 1962 an evaluation study was published [2] indicating the necessity for increasing the support of the theoretical research area as a means of enhancing the progress in this field. Figure 3 also describes the effect of this recommendation on the national effort in weather modification. In this case we wish to examine the curves labeled 1966, which interestingly enough cover the period during which the National Academy of Science [3] made essentially the same recommendation as was contained in Ref. 2. As can be seen, the research effort has actually lost ground during this time (considerably more than the difference between 2.71 million dollars and 2.5 million dollars because of the increase in costs between 1962 and 1966). On the other hand the experimental program has more than doubled (presumably under the pressure to do something now), thereby enhancing the observed dangerous imbalance in the total program.

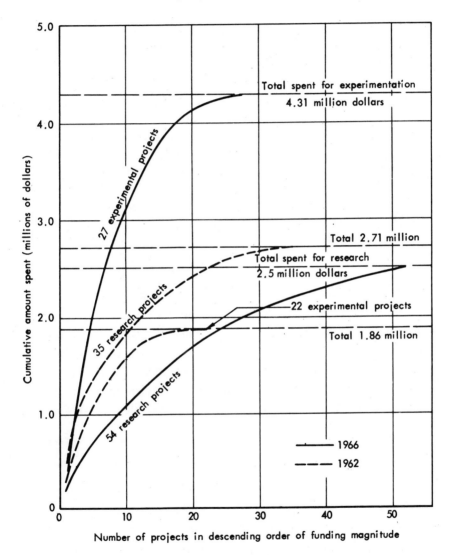

Figure 3.   Distribution of funding for weather modification.

The second additional reason lies in the fact that there does not exist a supported group in this country today charged with engaging in the *total problem of weather modification.** Such a group would provide the required third force that could bring researchers to interact across the artificial scale boundaries and that which appears to exist between the experimenter and the modeler. Such a group must be involved in the total subject, and be totally committed. Yet, for some unknown reason, possibly jurisdictional or political, no governmental agency today is willing to recognize the total problem for what it is, and face up to the necessity of accepting the challenge and supporting the program the way it should be supported. Until some agency accepts that challenge, this situation is not going to change. The weather modification effort may get larger in the terms of the total money that is being spent, but its character will not change and if we were to come back ten years from now, and look at the situation again, then in the absence of the acceptance of the total challenge, we would probably still have this dichotomy existing between experiment, research and scale size. Even more important, we would probably not know very much more about the total atmosphere, and our ability to modify it and control it, than we know today.

A final worrisome problem is contained in the fact that, despite the obvious magnitude of the need for weather modification effort, this area itself is only one part of the total challenge facing man today. The real problem concerns the ability of man to conserve, control, and manage the environment in which we all live. As such it extends across the problem of air pollu-

---

* **Editor's note:** A theme which has been reiterated in several ways during this conference has been the insight that one needs to consider the *total* system rather than a piece of it in order to understand it, design it, control it, etc. In this connection I would like to speculate that one of the greatest spin offs of the NASA mission will be the techniques they developed to create the management-machine system responsible for the successful moon landing. The is a unique case in which the *total problem* was attacked and completely coordinated. The enormous attention they gave to the use of feedback (even to monitoring all the important subsystems on the vehicle from the ground) stands in marked contrast to the lack of feedback existing in most of the other man-made large-scale systems that exist today whether they be machine or social systems.

tion, water pollution, thermal pollution and weather modification, and every other way man directly or indirectly insults his environment. It's concerned with the effects of cities and the effects of the growing demands for energy and electrical power. It's concerned with all of these effects interacting with one another in the physical, biological, economic and social sense. The challenge is a large one, because I am afraid they cannot be considered separately. There is a great temptation to recoil at the size of the problem and to retreat back into our own restricted areas of interest. I am afraid if one is truly concerned about the viable future of man in the presence of his environment, then one cannot allow this to happen. My impression of the choice facing us in attempting to decide or control man's fate in this connection is relatively simple. We can extrapolate into the future, accepting the directions provided by the present as fixed. In this case, you cannot control, nor are you assured of the desirability of your end point, because you cannot observe or handle all of the branch points that will occur. Alternatively, you can specify the future you desire, and then examine the available ways of getting there from the present.

In the case of man's environment we cannot afford to extrapolate into the future. The effort to determine how to reach a future we specify *demands that we attack the entire problem.*

REFERENCES

1. V. J. Schaefer and Irving Langmuir, *The Modifications Produced in Clouds of Super-Cooled Water Droplets by the Introduction of Ice Nuclei.* Paper presented at the Annual Meeting of the American Meteorological Society, Boston, Massachusetts, December 1946.
2. S. M. Greenfield, R. E. Huschke, Yale Mintz, R. R. Rapp, and J. D. Sartor, *A New Rational Approach to Weather Control Research.* RM-3205-NSF, the RAND Corporation, Santa Monica, California, May 1962. Reprinted with permission.
3. National Academy of Sciences-National Research Council, *Weather and Climate Modification Problems and Prospects, 1,* summary and Recommendations, 2. Research and Development, NAS-NRC, Publ. No. 1350, Final Report of the Panel on Weather and Climate Modification.
4. M. Ewing and W. L. Donn, "A Theory of Ice Ages." *Science, 123,* June 15, 1956, 1061-1066; *127,* May 16, 1958, 1159-1162.

5. Staff, The Weather Modification Research Project of The RAND
   Corporation, *Weather Modification Progress and the Need for Inter-
   active Research*. RM-5835-NSF, the RAND Corporation, October 1968.
   Reprinted with permission.

Index

261